FEBRUARY 2023 — VOL. 5

On the Horizon

A Collection of Papers from the Next Generation

EDITOR
Reja Younis
Jessica Link

AUTHORS
Jasmin Alsaied
Caitlyn Bess
Brandon Cortino
Daniel Gum
Annika Kastetter
Jaejoon Kim
Dilan Ezgi Koç

Lansing S. Horan IV
John Madeira
Aubrey Means
Wil Powell
Erica Symonds
Dan Zhukov

CSIS | CENTER FOR STRATEGIC &
INTERNATIONAL STUDIES

ROWMAN & LITTLEFIELD
Lanham • Boulder • New York • London

About CSIS

The Center for Strategic and International Studies (CSIS) is a bipartisan, nonprofit policy research organization dedicated to advancing practical ideas to address the world's greatest challenges.

Thomas J. Pritzker was named chairman of the CSIS Board of Trustees in 2015, succeeding former U.S. senator Sam Nunn (D-GA). Founded in 1962, CSIS is led by John J. Hamre, who has served as president and chief executive officer since 2000.

CSIS's purpose is to define the future of national security. We are guided by a distinct set of values—nonpartisanship, independent thought, innovative thinking, cross-disciplinary scholarship, integrity and professionalism, and talent development. CSIS's values work in concert toward the goal of making real-world impact.

CSIS scholars bring their policy expertise, judgment, and robust networks to their research, analysis, and recommendations. We organize conferences, publish, lecture, and make media appearances that aim to increase the knowledge, awareness, and salience of policy issues with relevant stakeholders and the interested public.

CSIS has impact when our research helps to inform the decisionmaking of key policymakers and the thinking of key influencers. We work toward a vision of a safer and more prosperous world.

CSIS does not take specific policy positions; accordingly, all views expressed herein should be understood to be solely those of the author(s).

ISBN: 978-1-5381-7064-9 (pb); 978-1-5381-7065-6 (ebook)

Center for Strategic & International Studies
1616 Rhode Island Avenue, NW
Washington, D.C. 20036
202-887-0200 | www.csis.org

Rowman & Littlefield
4501 Forbes Boulevard
Lanham, MD 20706
301-459-3366 | www.rowman.com

About the Project on Nuclear Issues

The Project on Nuclear Issues (PONI) was developed in 2003 to develop the next generation of policy, technical, and operational nuclear professionals by fostering, sustaining, and convening a networked community of emerging experts. PONI identifies and cultivates emerging thought leaders by building relationships, deepening understanding, and sharing perspectives across the full range of nuclear issues and communities. PONI's programs provide inclusive, diverse, and creative opportunities for rising experts to learn about policy, technical, and operational aspects of the nuclear community, develop and present new concepts and ideas, engage in thoughtful and informed debates, and tour and visit sites across the nuclear enterprise.

PONI strives to achieve this mission through several objectives:

- Identifying emerging thought leaders and providing them with the opportunity to develop and present new concepts and ideas.

- Sponsoring new cutting-edge research.

- Encouraging thoughtful and informed debate.

- Engaging a broad and diverse community across the country and internationally.

- Providing a networked platform for information-sharing and collaboration across the broad nuclear community.

- Cultivating young professionals through opportunities to build relationships, deepen understanding, and share perspectives across the full range of nuclear issues and communities.

PONI sponsors numerous opportunities for young professionals to engage in thoughtful and informed debate on the nuclear community's most pressing challenges.

PONI strives to expand its outreach to address all career and academic levels, connect young professionals in collaborative research projects, broaden the topics it covers across the full spectrum of nuclear issues, and ensure robust inclusion of expertise from all critical domains – academic, military, scientific, and technical.

1. Inclusivity – Welcome all ideas and perspectives across political, ideological and policy spectrum.

2. Diversity – Actively seek interdisciplinary perspectives (technical, operational, corporate, government, academic) and embrace participation across all demographics.

3. Creativity – Promote collaborative, innovative research and dynamic, engaging programming.

Amongst the various programming opportunities available through PONI, the authors in this publication were members of PONI's 2021 Nuclear Scholars Class. The PONI Nuclear Scholars Initiative is a group of select graduate students and young professionals. The Nuclear Scholars Initiative aims to provide top graduate students and young professionals from around the country with a unique venue to interact and dialogue with senior experts on nuclear weapons issues. Those accepted into the program are hosted monthly at CSIS in Washington, DC, where they participate in daylong workshops with senior government officials, policy experts, and technical experts. Over the course of the six-month program, Scholars are required to prepare a research product. PONI has several alumni from this initiative, many of which continue to work in the nuclear field and play key roles in nuclear policy development, technical innovations, and operations.

Acknowledgments

PONI owes many thanks to the authors for their dedication throughout the research and writing process, and outstanding products. We appreciate senior experts and Mid-Career Cadre members who reviewed and edited; provided mentorship and guidance; and came to speak with the Nuclear Scholars during their workshops. PONI could not function without their generosity and support. The editors also received tremendous support from their colleagues at PONI, including Nicholas Adamopoulos, Joseph Rodgers, and Genevieve Hackman, with proofreading and feedback.

PONI would also like to thank the CSIS External Relations and iLab, including Alex Kisling, Jeeah Lee, Katherine Stark, Rayna Salam, Phillip Meylan, and William Taylor for their help in the editing, graphic design, and publication of the report for their support in editing and publishing the report.

Lastly, PONI would like to express gratitude to our partners for their continued support. Amongst our many partners this publication was made possible with support by the National Nuclear Security Administration. This material is based upon work supported by the Department of Energy National Nuclear Security Administration under Award Number DE-NA0003970.

This report was prepared as an account of work sponsored by an agency of the United States government. Neither the U.S. government nor any agency thereof, nor any of their employees, makes any warranty, express or implied, or assumes any legal liability or responsibility for the accuracy, completeness, or usefulness of any information, apparatus, product, or process disclosed, or represents that its use would not infringe privately owned rights. Reference herein to any specific commercial product, process, or service by trade name, trademark, manufacturer, or otherwise does not necessarily constitute or imply its endorsement, recommendation, or favoring by the U.S. government or any agency thereof. The views and opinions of authors expressed herein do not necessarily state or reflect those of the U.S. government or any agency thereof.

Contents

Introduction

By Reja Younis

The past year has tested the nuclear community like never before. Russia's invasion of Ukraine carries on with no end in sight to the fighting that has killed thousands, uprooted millions, and reduced Ukrainian cities to rubble. In this watershed strategic moment—laden with multidomain complexity, emerging technologies, and heated nuclear rhetoric—it is evident that the Russian invasion has brought war back to Europe. But the Russian invasion has also done something else: it has reinvigorated a nuclear intellectual renaissance. Deep expertise and passion will be critical to meet the nuclear challenges of the present and future. This will require a sustained long-term effort by a multidisciplinary force of nuclear experts who are equipped with critical knowledge, skills, and a robust professional network. It will also require new, creative, and diverse thinking on how to approach various political, military, legal, ethical, and technical challenges in the United States and around the world.

Recognizing this challenge, the Center for Strategic and International Studies (CSIS) launched the Project on Nuclear Issues (PONI) in 2003 to develop the next generation of policy, technical, and operational nuclear professionals by fostering, sustaining, and convening a networked community of emerging experts. PONI seeks to revitalize and strengthen the community of nuclear experts whose training and background increasingly emphasize multidisciplinary expertise, especially among young generations. PONI runs two signature programs for young professionals—the Nuclear Scholars Initiative and the Annual Conference Series—to engage rising nuclear experts in thoughtful and informed debate on how to best address the nuclear community's most critical issues. This volume is comprised of papers from participants in the 2022 Nuclear Scholars Initiative. PONI sponsors this research to provide a forum for facilitating new and innovative thinking and providing a platform for fresh thought leaders across the nuclear enterprise. Through a process of peer review, mid-career mentorship, and senior expert review, nuclear scholars are encouraged to immerse themselves in tough questions and pursue creative solutions. The papers in this volume span a wide range of policy and technical issues, further discussion in their respective areas, and provide innovative recommendations for current challenges. To that end, these papers explore pressing topics such as the AUKUS security pact, gray zone security assurance, the impact of climate change on nuclear stability, deterring two near-peer adversaries, cyber nuclear diplomacy, and small modular reactors. There is something in this volume for everyone in the nuclear community. That is PONI's mission.

The AUKUS Security Pact

A New Precedent for Non-Nuclear Weapons States and Naval Nuclear Propulsion

By Jasmin Alsaied[1]

ABSTRACT

In 2016, the Australian government's Department of Defence published a report detailing Australia's desire to acquire a submarine fleet.[2] The report cited the necessity to "mount a sustained presence . . . [throughout the] many straits and geographical features that form choke points [throughout Southeast Asia]." The report also argued that Australian maritime strategy, the country's economic prosperity, and the welfare of its citizens and allies could all be strengthened by Australia's development of an undersea presence. The feasibility of use, the extended range at which submarines can operate, and the notion of a submarine acting as a deterrent to adversaries are all cited as tangible reasons to seek and pursue the technology.[3] The Australian government acknowledges the potential prowess and numerous advantages given by nuclear propulsion technologies. At the time of the report's publication, however, Australia's pursuit of naval nuclear propulsion was at a standstill because the country lacked the necessary information, technology, infrastructure, and resources to develop a nuclear fleet, develop the necessary support systems, and train crews to operate and maintain submarines. Six years later, national security considerations have compelled the Australian government to seek robust and dependable technologies that can power a fleet of attack submarines to protect national assets and interests. Those technologies will be delivered in the newly agreed upon AUKUS security pact.

---

1   LT Jasmin Alsaied is currently a U.S. Navy surface warfare officer. The views expressed in this paper are those of the author and do not reflect the official policy or position of the Department of the Navy, the Department of Defense, or the U.S. government.

2   Janis Cocking, Chris Davis, and Christopher Norwood, *Australia's requirement for submarines* (Canberra: Australian Department of Defence, Science and Technology, 2016), https://www.dst.defence.gov.au/sites/default/files/publications/documents/DSC%201618%20-%20DST%20Submarine%20Report%20PRO4%20LR.pdf.

3   Ibid.

With the enaction of the AUKUS security pact between the United States, United Kingdom, and Australia, the country will possess the necessary technology and know-how to operate a nuclear-powered submarine fleet. This is only the second time the United States has shared this technology, providing it to the United Kingdom in 1958.[4] An abiding signatory of the Treaty on Non-Proliferation of Nuclear Weapons (NPT), Australia will forge a path as the first non-nuclear weapons state (NNWS) to independently operate and maintain nuclear-propelled submarines. Australia could set a precedent as an NNWS should they successfully implement new safeguards and guidelines to becoming an independent naval nuclear propulsion operator. This paper aims to explain the challenges, changes, and precedent Australia must both overcome and newly set in order to successfully obtain naval nuclear assets as an NNWS.

An in-depth examination of Australia's current relationship with and safeguard regime under the International Atomic Energy Agency (IAEA) will be examined below to determine what changes need to be made in Australia's current methodology and technology-sharing processes in order to remain a firm adherent to the NPT while obtaining this new technology. Aspirational countries such as Australia who are looking to implement naval nuclear technologies are unlikely to covertly pursue proliferation regimes to develop malicious nuclear weapons. Australia has shown its commitment to nuclear safeguards by maintaining a robust safeguards regime and pursuing several encompassing treaties with other Asian partners to minimize the proliferation of nuclear weapons programs in the region.

The intent of this paper is not to explore Australia's motive or ability to evade safeguards and pursue nuclear weapons capabilities but instead to outline and highlight the challenges the nation will face in adhering to current safeguards while pursuing naval nuclear propulsion technology. Australia's successful acquisition of naval nuclear propulsion technology could provide a framework for other trusted allies to join the naval nuclear regime within the Indo-Pacific region, without weakening the existing nonproliferation status quo. The AUKUS pact's successful delivery of nuclear-powered submarines could demonstrate that nonproliferation norms can be upheld in cases where a receiving country avoids the fabrication and assembly of the reactor core.

THE AUKUS SECURITY DEAL AND NUCLEAR SUBMARINES: A PRIMER

On September 15, 2021, the Biden administration announced a new strategic security pact that set out to deepen defense ties between the United States, United Kingdom, and Australia in naval, cyber, artificial intelligence (AI), quantum computing, and other undersea domains.[5] The pact, dubbed "AUKUS," creates a platform to provide an increased presence and enhanced force posture in and around Australia in response to growing threats from malign Chinese influence and technologies in the region. Other military capabilities will be improved through robust data-sharing regimes and bolstered by an elevated U.S. force posture. The deal initially garnered media attention when it was revealed that Australia, an NNWS, would begin constructing and operating at least eight submarines powered by nuclear propulsion.[6] The submarines, powered by highly enriched uranium (HEU), have already sparked great

---

4   Ayesha Rascoe and Alana Wise, "Why Biden Is Taking the Rare Step of Sharing Nuclear Submarine Tech with Australia," NPR, September 16, 2021, https://www.npr.org/2021/09/15/1037338887/why-biden-is-taking-the-rare-step-of-sharing-nuclear-submarine-tech-with-austral.

5   "Remarks by President Biden, Prime Minister Morrison of Australia, and Prime Minister Johnson of the United Kingdom Announcing the Creation of Aukus," (speech, The White House, September 15, 2021), https://www.whitehouse.gov/briefing-room/speeches-remarks/2021/09/15/remarks-by-president-biden-prime-minister-morrison-of-australia-and-prime-minister-johnson-of-the-united-kingdom-announcing-the-creation-of-aukus/.

6   Ibid.

controversy with allies such as France and adversaries such as China. Critics also expressed concern over the potential for another country to design and maintain nuclear weapons under the precedent Australia may set in obtaining nuclear materials and technology.[7] The leaders of the three countries, however, emphasized that the AUKUS security pact is meant to share technology only intended for use in naval nuclear propulsion. During an announcement regarding the AUKUS deal, President Joe Biden emphasized the following: "I want to be exceedingly clear about this: we're not talking about nuclear-armed submarines. These are conventionally armed submarines that are powered by nuclear reactors."[8] When the first shipments of nuclear-operated submarines are delivered to Australia in approximately a decade's time, the country will become the first NNWS to possess naval nuclear propulsion technology.[9]

Nuclear propulsion capabilities have dramatically improved since the emergence of the first nuclear submarine in 1955.[10] At the time, nuclear cores could endure for only 62,000 miles.[11] Today, it is estimated that nuclear cores can operate for over 1,000,000 miles.[12] These improved capabilities allow submarines to maintain an element of stealth and remain operational for longer periods of time, only surfacing or returning to port for crew needs. Though the appeal of nuclear submarines is fairly evident, one concern with naval nuclear propulsion is its requirement of HEU, which creates inherent proliferation concerns due to its dual-use capability in nuclear weapons. Nuclear cores produced in the United States and United Kingdom, both NWSs, utilize over 90 percent enriched uranium and provide consistent, efficient, and reliable outputs of energy for shipboard use.[13] The Australian ambassador to the United States confirmed that the AUKUS security pact will utilize HEU to avoid the need for a civilian nuclear industry.[14] The composition of this intended fuel choice provides flexibility in maintaining IAEA safeguards but still poses risks for nuclear critics who are concerned about proliferation risks surrounding the AUKUS defense pact. The largest proliferation risk surrounding an HEU-fueled submarine would be during the initial core construction and eventual return to supplier for core repair, refabrication, and replacement. Because the texts prefacing and guiding the AUKUS security pact and its particulars have either not yet been worked out or are not yet released, it is unclear if the construction of these nuclear cores will use existing technology or rely on new methods of export to supply HEU for Australia's core construction.

In this process, the IAEA should be responsible for addressing its role regarding the legal and technical challenges that will arise from the AUKUS security pact. AUKUS partners must be forthright

---

7   Rich Abott, "Ambassador: Aukus Subs to Avoid Domestic Nuke Industry, Confirms Heu," *Defense Daily*, November 17, 2021, https://www.defensedaily.com/ambassador-aukus-subs-to-avoid-domestic-nuke-industry-confirms-heu/navy-usmc/; and Bruno Tertrais, "France, America and the Indo-Pacific after AUKUS," Institut Montaigne, September 22, 2021, https://www.institutmontaigne.org/en/blog/france-america-and-indo-pacific-after-aukus.

8   Rascoe and Wise, "Why Biden Is Taking the Rare Step of Sharing Nuclear Submarine Tech with Australia."

9   This term refers to a nation that does not own or operate any nuclear weapons and the state of fissile material having been enriched to weapons-grade material for potential weapons production or nuclear propulsion use.

10   "Nuclear-Powered Ships," World Nuclear Association, November 2021, https://world-nuclear.org/information-library/non-power-nuclear-applications/transport/nuclear-powered-ships.aspx.

11   Department of Energy and the Department of the Navy, *The United States Nuclear Propulsion Program* (Washington, DC: 2020), https://www.energy.gov/sites/default/files/2021-07/2020%20United%20States%20Naval%20Nuclear%20Propulsion%20Program%20v3.pdf.

12   Ibid.

13   The NWS term refers to a nation that owns or operates nuclear weapons and fissile material having been enriched to weapons-grade material for potential weapons production or nuclear propulsion use. Frank N. Von Hippel, "U.S. Shift Away from HEU-Fueled Naval Nuclear Reactors Could Begin in the 2040s," International Panel on Fissile Materials, June 26, 2019, https://fissilematerials.org/blog/2019/06/us_shift_away_from_heu-fu.html; and Dan McMorrow, *Low-Enriched Uranium for Potential Naval Propulsion Applications* (McClean, VA: MITRE Corporation, November 2016), https://irp.fas.org/agency/dod/jason/leu-naval.pdf.

14   Rich Abott, "Ambassador: Aukus Subs to Avoid Domestic Nuke Industry, Confirms Heu," *Defense Daily*, November 17, 2021, https://www.defensedaily.com/ambassador-aukus-subs-to-avoid-domestic-nuke-industry-confirms-heu/navy-usmc/.

and take on the responsibility of clarifying and detailing all aspects of the transport, export, withdrawal, and shipment of nuclear material to partner states.

Failure to do so may substantiate the arguments of critics and states who see Australia's nuclear submarine acquisitions as a mechanism to propel a "proliferation-permissive" environment. Conversations regarding the precedent that Australia would set regarding the use of special nuclear material for peaceful military nuclear use are important and grounded. The pact could set the tone regarding how other states pursue nuclear-propelled submarine technology. Should the AUKUS pact be emulated among other countries, vested stakeholders must wrestle with the example that is set for other countries to parallel and exact in their own pursuits of nuclear-powered submarine technology. Further, prescriptions in this paper aim to explore feasible outcomes for the AUKUS agreement that could also be applied to other [n]-country portfolios that seek to uphold nonproliferation norms in pursuit of peaceful military use of nuclear material.

The following sections outline the current norms and status quo set by IAEA safeguards requirements and how Australia fits into that framework. The subsequent sections will then postulate how Australia's safeguards regime—and by extension those of other AUKUS partners—will need to flex and adapt to securely and efficiently provide Australia with a new fleet of nuclear-powered submarines.

THE INTERNATIONAL ATOMIC ENERGY AGENCY'S CURRENT VERIFICATION REGIME

IAEA STANDING GUIDANCE

As an NNWS party to the NPT, Australia falls under Information Circular (INFCIRC)/153, which governs all IAEA member states' comprehensive safeguards agreements (CSAs).[15] According to the IAEA, an NNWS can utilize HEU, or special nuclear material, for naval nuclear propulsion and is exempted from safeguards.[16] Any state that is party to these CSAs must inform the IAEA when importing or exporting any nuclear material containing thorium or uranium to an NNWS unless the transfer is for "specifically non-nuclear purposes."[17] INFCIRC/153 states that "shipments of yellowcake by the state concerned to a NWS or a NNWS for the production of fuel for a submarine reactor would not be excluded from such an obligation, since that is a non-nuclear purpose."[18] A state has the potential to be subjected to safeguards simply by engaging in "the production or import of nuclear material, regardless of the ultimate intended use, upon its entry into the state."[19]

Specifically, INFCIRC/153 states that safeguards should begin once material in the nuclear fuel cycle reaches a "composition and purity suitable for fuel fabrication."[20] Paragraph 14 of INFCIRC/153, however, accommodates countries who wish to utilize nuclear material in a "non-prescribed" military nuclear activity. INFCIRC/153 does not explicitly define "peaceful nuclear activity." For the purposes of

---

15 "The Structure and Content of Agreements between the Agency and States Required in Connection with the Treaty on the Non-Proliferation of Nuclear Weapons, IAEA document INFCIRC/153 (Corrected), 1972," International Atomic Energy Agency (IAEA), June 1972, https://www.iaea.org/sites/default/files/publications/documents/infcircs/1972/infcirc153.pdf. "INFCIRC" refers to an "Information Circular" document of the IAEA and designates a document that has been circulated to all member states. INFCIRCs are publicly available on the IAEA's website: http://iaea.org.

16 The Atomic Energy Act and 10 C.F.R. Part 810.3 define special nuclear material as enriched uranium, uranium-233, or plutonium. "INFCIRC/153(Corrected)," IAEA.

17 Ibid.

18 Ibid.

19 Ibid.

20 Ibid.

this paper, the definition used by the Federation of American Scientists will be utilized, which defines propulsion, transport, reprocessing, and storage as peaceful military nuclear activity, even when utilizing previously withdrawn nuclear material.[21] Thus, any state looking to withdraw this nuclear material for enrichment and fuel fabrication must first coordinate with the IAEA. The IAEA requires the state utilizing nuclear material for purposes that fall under Paragraph 14 to ensure that:

- the nonprescribed military activity will not conflict with an undertaking the state may be involved in which requires IAEA safeguards;
- nuclear material will only be used in peaceful nuclear activities; and
- nuclear material will not be used for the nuclear weapons or explosive devices.

To engage in any of the prescriptions outlined in Paragraph 14, the state must submit an arrangement for approval from the IAEA Board of Governors detailing the total quantity, composition, and circumstances of the nuclear material that will be withdrawn from safeguards. The Board of Governors is one of two policymaking bodies within the IAEA that approves safeguards agreements. The board is comprised of states who do and do not possess nuclear weapons and includes all NWSs recognized by the IAEA.[22] INFCIRC/153 states that the IAEA's agreement "shall not provide any approval or classified knowledge of the military activity or relate to the use of nuclear materials therein."[23] Because the IAEA is not privy to any state's classified information, the burden of the state is to only provide information to the extent possible in order to meet these information requirements.

NNWSs who are NPT signatories must maintain and provide detailed information regarding the control of this nuclear material. The IAEA can then verify the physical measurements and amounts via a special, routine, or ad hoc inspection. This governs a great deal of the nuclear material transfer that will need to occur to successfully facilitate the construction of Australia's new propulsion cores. Upon reintroduction of the withdrawn nuclear material back into peaceful nuclear activity (in this case, propulsion), safeguards are also reimplemented on the activity. The state will then continue to adhere to any information or inspection requirements delineated by the IAEA.

Arguments surrounding information regarding safeguards for peaceful nuclear military activity consider the standing guidance to be vague or to leave certain clauses as subject to interpretation. For the purposes of this analysis, this paper assumes that the states involved in the AUKUS security pact intend to invoke Paragraph 14 and come to an agreement with the IAEA prior to any potential nuclear material withdrawal for transport, enrichment, fabrication, reprocessing, or storage.

CASE STUDY: THE CANADIAN REQUEST

The nuclear technology sharing process implied within the AUKUS security pact, while novel in its information sharing for Australia, has been seen in the past in Canada's request to withdraw nuclear material from IAEA safeguards to use in nuclear-powered submarines. In the late 1980s, Canada proposed exporting and enriching uranium hexafluoride (UF_6), fabricating fuel elements, and returning the material for use in a new fleet of Canadian nuclear-powered submarines.[24]

---

21   Laura Rockwood, "Naval Nuclear Propulsion and IAEA Safeguards," Federation of American Scientists, Naval Nuclear Propulsion Project, August 2017, https://uploads.fas.org/media/Naval-Nuclear-Propulsion-and-IAEA-Safeguards.pdf.
22   "Board of Governors," IAEA, June 8, 2016, https://www.iaea.org/about/governance/board-of-governors.
23   "INFCIRC/153(Corrected)," IAEA.
24   Rockwood, "Naval Nuclear Propulsion and IAEA Safeguards."

To accommodate the IAEA's request for notification of the withdrawal of nuclear material, Canadian officials suggested withdrawing nuclear material in Canada and then subsequently exporting it to an NWS for enrichment and fabrication. The fuel elements would then be sent back to Canada for arrangement and use in submarines.[25] The IAEA's Secretariat concluded that the withdrawal of the nuclear material should occur at the latest stage possible to avoid any material being "lost to export" and to limit the time period that material would not be in use or under the custody of the Canadian government.[26] Canadian officials also proposed not invoking Paragraph 14 requirements and solely pursuing a direct military-to-military arrangement.[27] The IAEA Secretariat, however, likely acknowledged the compromising potential this would have on IAEA safeguards and regime verification and thus recommended an alternative approach that was eventually agreed to by all parties.

In response, Tariq Rauf and Marie Desjardins published a piece in 1988 titled "Opening Pandora's Box? Nuclear Powered Submarines and the Spread of Nuclear Weapons."[28] Their paper acknowledged the fact that Canada was intended to be the first NNWS that would engage in such enrichment and fabrication for naval nuclear propulsion and expressed great concern for the nuclear material that was planned to be outside of the IAEA's scope of safeguards. The authors argued that allowing Canada to export nuclear material for use in naval nuclear propulsion, though widely regarded as a peaceful practice, would open a Pandora's Box, and set a precedent for other rogue nations to invoke Paragraph 14 for illicit or subversive aims.

Reactions to this piece produced valid discussion points that can easily be applied to the Australian case. The proposal put forth by AUKUS partners will likely need to address these concerns to successfully promote a safe technology and equipment transfer that is minimally susceptible to export losses or vague safeguards limitations. Though the Rauf and Desjardins paper acknowledges the use of nuclear "guardrails" to promote accountability and strength in safeguards, the authors held apparently little faith in the coordinated technical and advisory capabilities of the IAEA.

Canada eventually chose not to pursue the program, but this case study shows that an arrangement between the United States, United Kingdom, and Australia has the potential to conform to the IAEA's guidelines, maintain the necessary classification of naval nuclear technology, and effectively limit the time that nuclear material would be removed for enrichment and fabrication. The current tripartite agreement could likely draw many similarities to Canada's proposed plan and develop further precautions to safeguard information and technology while pursuing a robust plan to incorporate Australia into the naval nuclear regime.

AUSTRALIA'S SAFEGUARDS REGIME

Australia maintains a close relationship with the IAEA and has stood as a strong proponent of safeguards, ratification of the NPT, and the nuclear nonproliferation regime for several decades. Australia is also home to some of the largest natural deposits of uranium.[29] As such, the country

---

25   Ibid.
26   Ibid.
27   Ibid.
28   Tariq Rauf and Marie-France Desjardins, "Opening Pandora's Box? Nuclear-Powered Submarines and the Spread of Nuclear Weapons," Canadian Centre for Arms Control and Disarmament, *Aurora Papers* 8 (March 1988), https://www.researchgate.net/publication/350017528_Opening_Pandora's_Box_Nuclear_Powered_Submarines_and_the_Spread_of_Nuclear_Weapons_The_Canadian_Centre_for_Arms_Control_and_Disarmament_Aurora_Papers_8_1988.
29   "Australia's Uranium," World Nuclear Association, April 2022, https://world-nuclear.org/information-library/country-pro-

upholds and advertises a comprehensive safeguards agreement and imposes bilateral safeguards on all Australian uranium exports. Australia has consistently been lauded as one of the best NPT signatories by the Nuclear Threat Initiative for its contributions to continuing efforts to strengthen nuclear security.[30] In short, Australia is a model country to emulate regarding safeguards imposition, nuclear security, treaty ratification, and enforcement. The true test of AUKUS partners will be determining if Australia can maintain this status under the new deal. The following sections outline potential changes to Australia's current safeguards regime and compliance structure that AUKUS partners will likely need to address in order to abide by IAEA guidelines and Australia's own stringent safeguards regime.

Australia can expect to make robust changes to its current verification regime to successfully obtain, utilize, and maintain a nuclear-propelled submarine fleet. The greatest issue that Australia faces is ensuring that the details prescribed in the agreement postulated by the three countries addresses the IAEA's concerns over the ambiguous directives governing nuclear material withdrawal for peaceful military activity. The burden of proof will likely fall on the AUKUS signatories to prove that their proposed method of enrichment, transport, and fabrication will minimize any chances of material being "lost to export" or falling prey to technology manipulation, theft, or tampering. Proposals made will likely need to discuss the quantity and composition of withdrawn nuclear material, the timeline for withdrawal and reintroduction to safeguards, details regarding shipment and export schemes, and other various reporting requirements. Great care must be taken when drafting the AUKUS agreement with the IAEA; Paragraph 14 of INFCIRC/153 does not state the level of detail required in the reporting of withdrawal of nuclear material. AUKUS partners should prepare for numerous internal, trilateral, and potentially quadrilateral conversations regarding the information required and level of detail presented to the IAEA regarding an invocation of Paragraph 14.

ACCOUNTING FOR LOSSES

According to the IAEA, all states subject to CSAs are required to implement safeguards to ensure that nuclear material that is withdrawn for enrichment and fabrication purposes is not also or instead diverted for use in nuclear weapons. The IAEA has a corresponding "right and obligation to ensure that safeguards will be applied, in accordance with the terms of the agreements on all source and special fissionable material in all peaceful nuclear activity within its territory, under its jurisdiction or carried out under its control anywhere."[31] These requirements are bound to create issues in oversight and application.

Per the IAEA's current safeguards system, any arrangement designed to govern AUKUS construction activities should include precautions to limit the potential for withdrawn materials to be inaccessible to safeguards through exports. In the proposed agreement made by AUKUS partners, specific verbiage needs to be dedicated to the adequate implementation of safeguards that will prevent any inadvertent loss of material during the enrichment and fabrication process. Thus, the three states involved in the construction of the nuclear cores will need to maintain records for the physical accounting of the material removed from safeguards for enrichment, fabrication, and final use in the construction of the nuclear core. The trilateral partnership should also provide new approaches or guarantees to safeguards in order to overcome potential challenges in protecting material from loss to export.

files/countries-a-f/australia.aspx.

30   "Australia Overview," Nuclear Threat Initiative, October 19, 2021, https://www.nti.org/analysis/articles/australia-overview/.

31   "The Statute of the IAEA," IAEA, June 2, 2014, https://www.iaea.org/about/statute.

Paragraph 14 of INFCIRC/153 also clearly states that the IAEA has no right to classified information. However, Paragraph 14 explicitly requires that the IAEA be "kept informed of" the total quantity and composition of such unsafeguarded nuclear material in the state and of any exports of such material.[32] There is some precedent regarding the discussion of classified or sensitive information that the IAEA may be privy to. The IAEA could agree to receive ancillary documents and protect them from distribution should the state deem them confidential or to concern safeguards on confidential material in a potential agreement with AUKUS partners.[33] The IAEA Board of Governors also reserves the right to take appropriate action in each case and could deem the circumstance of AUKUS one that can be supported and presented to the board via Subsidiary Arrangements that are protected from distribution. This arrangement, between the state and the IAEA, is not approved by the board but is normally protected as safeguards confidential. Doing so would protect the integrity of the IAEA's safeguards, even when excepted under Paragraph 14, and would also protect the integrity of naval nuclear propulsion technology and data of the United States, United Kingdom, and Australia.

Paragraph 14, however, does not mandate the IAEA's right to verify via inspection or any means. Further, INFCIRC/66/Revision 2 states that non-signatories to the NPT will allow the IAEA to certify that nuclear materials or items placed under its supervision are "not being used in any prohibited military activity."[34] Under this directive, the IAEA is unable to give assurances regarding the production of nuclear weapons or comment on the state of the country's nuclear activities or pursuits. This clause, however, is only utilized by non-signatories of the NPT. Australia could try to use this verbiage to set precedent in its proposal and request that the IAEA refrain from commenting on the state of its nuclear propulsion pursuits or activities. Australia can grant a supervisory status to the IAEA as detailed in INFCIRC/66 and submit to routine verification at certain stages in the material withdrawal process, on the condition that the IAEA would not comment on the nature or status of the activity taking place in order to protect the integrity of the information provided.[35]

Additionally, the IAEA requires the identity of the state that is importing or exporting nuclear material to be known. Since the IAEA is not privy to classified information, it will be difficult for the involved parties to subject themselves to an adequate level of safeguards during transport, reprocessing, or storage. Limited safeguards may need to be approved by the IAEA Secretariat and involved parties as a way of ensuring that the IAEA is kept informed and appraised throughout the withdrawal process while also maintaining a high level of confidentiality. So how do the constraints of classification, fuel, and other politics actually play out in feasible scenarios?

AUKUS COURSES OF ACTION

This paper lays out three potential courses of action that take account for the constraints and limitations set forth by the IAEA and AUKUS partners.

Option 1: Minimize Australia's involvement in the construction and acquisition of the nuclear submarine and invoke a Paragraph 14 exemption with the IAEA. Use limited safeguards and open dialogue to quell nonproliferation concerns. (This is a more likely option.)

---

32   "INFCIRC/153(Corrected)," IAEA.
33   Ibid.
34   "INFCIRC/66/Rev.2 - The Agency's Safeguard System," IAEA, 1968, https://www.iaea.org/sites/default/files/publications/documents/infcircs/1965/infcirc66r2.pdf.
35   Ibid.

Simply put, Australia could instead lease nuclear-powered submarines from other AUKUS partners. Alternatively, Australia could receive sealed nuclear reactor cores and only participate in the overall construction of domestic shipyard infrastructure to support building the hull and assembling the entire submarine. In this construct, the United States and United Kingdom, as skilled submarine outfitters and parties in the Nuclear Suppliers Group, would largely take on the burden of the construction and delivery of the submarine. Australia, then, would only be responsible for preparing for the receipt and subsequent training needed to operate, maintain, and service the submarines. The burden of ensuring that there is no loss to transport or export would still remain a priority of AUKUS partners. AUKUS partners could pursue this option because it does not threaten Australia's stance on nonproliferation and does not require any alterations to Australian federal law.

There is some precedent for this course of action. India leased a ballistic missile submarine in 1988 from the former Soviet Union and is seeking to lease another Akula-class nuclear-powered attack submarine for a hefty $3 billion from the Russian Federation.[36] This construct allows India to forego the troubles and complications of construction and delivery. Furthermore, India does not have to deal with the issues of safeguards on new or spent fuel and simply must participate in the training, maintaining, and operating of the nuclear submarine. Without speculating on India's intent to pursue a domestic nuclear submarine program as a non-signatory to the NPT, this construct relieves the receiving entity from the burdens associated with nuclear material and leaves less to be speculated upon regarding the location, transport, and use of enriched uranium.

Under the AUKUS security pact, Australia, which has already expressed a disinterest in indigenous enrichment schemes or core construction processes, could simply receive nuclear reactors or await the delivery of the nuclear-powered submarines and solely engage in training on the operation, servicing, and maintaining of the boat. Because of Australia's stance on the NPT, nuclear material, and safeguards application, this course of action may be the most attractive to the Australian government and Royal Australian Navy.

SAFEGUARDS APPLICATION

This course of action implies that AUKUS signatories will invoke a Paragraph 14 exemption with the IAEA and make the case that the submarines' fuel receipt, delivery, and construction fall under peaceful military nuclear use. Limited safeguards would apply during delivery and construction, as dictated during sessions between the IAEA and AUKUS partners, and could rely on alternative methods as listed above to meet information requirements for the location and use of special nuclear material. However, the United States and United Kingdom could rely on their past actions with the IAEA and invoke safeguards as needed during specific periods during construction of the nuclear reactors. This means that the IAEA could apply safeguards without concerns about the vulnerability of sensitive core designs. At the end of the submarine's service life, the spent fuel would return to the United States or United Kingdom and be dealt with accordingly. Spent fuel could then be subjected to safeguards upon arrival at storage facilities. Australia would not engage in the construction of any nuclear fuel facilities, enrichment facilities, or reprocessing facilities in support of naval nuclear propulsion. To reiterate, this option relies on Australia only acquisitioning the national technical means necessary to operate, service, and maintain the submarines.

---

36   Peter Suciu, "Why India Loves To 'Rent' Russian Nuclear Submarines," 1945, May 26, 2022, https://www.19fortyfive.com/2022/05/why-india-loves-to-rent-russian-nuclear-submarines/.

The IAEA's advisory services can support the AUKUS defense pact's petition through this course of action with the use of an Additional Protocol. In conjunction with CSAs, the protocol provides additional tools for verification to ensure that member states maintain peaceful uses of all nuclear material.[37] Those working to defend the AUKUS pact can utilize the protocol to provide the IAEA expanded access rights to information to ensure a continual verification of nuclear material. According to an IAEA informational website, "the Additional Protocol increases the IAEA's ability to provide much greater assurance on the absence of undeclared nuclear material and activities [within] States."[38] The AUKUS pact could strengthen the effectiveness of IAEA safeguards by allowing additional verification measures to generate greater assurances to IAEA member states and regional partners. With an emphasis on the care and nature of the information provided, the protocol allows the IAEA to fill in any gaps in the information provided in the initial proposal that the Board of Governors may feel is necessary to the safe use and implementation of Paragraph 14 exemptions. Notably, the United States, United Kingdom, and Australia have all adopted the protocol.[39]

NONPROLIFERATION CONCERNS

This course of action does nothing to address the nonapplication of safeguards during the delivery and construction of modern nuclear submarines. It does, however, ease concerns of Australia using the spent fuel, technical designs, equipment, and other national technical means of pursuing and developing a nuclear weapons program.

Option 2: Australia takes a larger role in the construction and delivery of nuclear submarines and invokes a Paragraph 14 exemption with the IAEA. Limited safeguards are applied. (This is the least likely option.)

This course of action implies that Australia may seek to pursue unique reprocessing or fabrication capabilities domestically to easily service their new nuclear-operated submarine fleet. Australia could also seek to store spent fuel at the end of service life within the country. In this construct, Australia would have to alter federal law to engage in the export, processing, and enrichment of its own nuclear fuel. This is likely the least attractive option to AUKUS partners, as it would inflate costs and the acquisition timeline and create domestic and international concerns regarding nuclear material transport and enrichment.

SAFEGUARDS APPLICATION

Should Australia choose to pursue any nuclear reactor construction, the IAEA must be immediately informed, and facilities would be subject to applicable safeguards. Should Australia develop the infrastructure to store spent fuel, the AUKUS pact would need to inform the IAEA of the material and quantity and place the material under safeguards once removed from reactor cores. Similar to the first course of action, limited safeguards would still be applied during the initial constructing and servicing of the nuclear reactors.

NONPROLIFERATION CONCERNS

Unlike Canada, Australia lacks the industrial, financial, or technical capability to build, operate, or maintain a gas-centrifuge plant to enrich uranium. Unless AUKUS partners pursue the development of an enrichment infrastructure within Australia, the enriched uranium to be used in propulsion must

---

37   "Additional Protocol," IAEA, June 8, 2016, https://www.iaea.org/topics/additional-protocol.

38   Ibid.

39   "Status List - Conclusion of Additional Protocols," IAEA, December 31, 2021, https://www.iaea.org/sites/default/files/20/01/sg-ap-status.pdf.

be imported from a trusted source. An important challenge here will be addressing Australia's federal law, which prohibits any uranium enrichment within the country. Australia would either need to alter federal law to pursue enrichment opportunities within the country or obtain special permission in its proposal to the IAEA Secretariat to pursue enrichment in safe facilities abroad (likely within the United States, United Kingdom, or other members of the Nuclear Suppliers Group). This construct would need to guarantee safe and expedited transport to the core construction location. This could complicate procedural compliance while nuclear material is excepted under Paragraph 14.

If Australia were to export natural uranium, it may not be subjected to safeguards under INFCIRC/217. Conversely, natural uranium that was processed prior to its shipment for enrichment would become subject to IAEA safeguards in the recipient's state. Further, details regarding uranium conversion, enrichment, and fabrication remain unclear and would pose additional risks depending on the process for fuel enrichment. CANDU's heavy-water moderators were a less risky option to enrich fuel, but Australia's prospects in fuel enrichment are lackluster at best and remain unclear. However, if the intent is for Australia to eventually semi-independently operate, maintain, and repair their nuclear fleet, the country may be forced to adopt an indigenous infrastructure to sustain submarine operations at varying operational tempos.

Another detail that should be accounted for in the AUKUS arrangement is spent fuel management. Australia is not currently equipped, through infrastructure or technical means, to handle spent fuel at the end of a submarine's lifetime. The technical expertise and support of the United States and United Kingdom will be vital in addressing concerns regarding spent fuel management at the conclusion of Australia's nuclear submarine life cycle. Should Australia find the challenges of storing spent fuel too great, per precedent set by the U.S. nuclear submarine program, the AUKUS pact could exercise the option of Australia returning all spent fuel to the United States or United Kingdom for subsequent storage and protection.

Again, should fuel be enriched in a different location than core construction, additional safeguards or measures would likely need to be detailed to AUKUS partners and the IAEA Secretariat to ensure that nuclear material is not lost due to export. Should core construction be largely handled by the United States and United Kingdom, this would not contravene Australia's federal law or current nuclear safeguards regime. Ultimately, the specific composition and location of the nuclear material in question distinguishes if and when Paragraph 14 must be invoked. Sensitive information regarding the enrichment and export scheme will likely not be released due to the risks it poses to potential proliferators and potential classification issues; thus, this section of the agreement must remain speculatory to outside observers and is likely a sensitive part of the dialogue between AUKUS partners and the IAEA Board of Governors.

Option 3: Direct military-to-military transfer of nuclear material for submarine core construction. AUKUS does not invoke a Paragraph 14 exemption. (This is a less likely option.)

Though undermining the capacity of the IAEA, utilizing a direct military-to-military transfer is still an option that AUKUS parties retain should they find Paragraph 14 and the IAEA's requests too restraining regarding the withdrawal and transfer of nuclear material. Such an arrangement could conceivably take place with the status of the Subsidiary Arrangements discussed. Given that all states parties consistently uphold international norms, it is unlikely that either the United States, United Kingdom, or Australia will try to legally invoke this pretext or move forward without some sort of agreement in place with the IAEA.

The Federation of American Scientists argues that AUKUS partners do not legally require the approval of the IAEA Board of Directors to enact the prescriptions of Paragraph 14.[40] Though the Canadian government proposed the idea of conducting a military-to-military transfer and continuing without IAEA approval, Canada's petition for a nuclear submarine fleet eventually opted to invoke Paragraph 14 and draft a proposal that was approved by the IAEA. No other member state of the IAEA has successfully invoked Paragraph 14 and conducted peaceful nuclear military activity in this context, nor has any member state chosen not to invoke Paragraph 14 regarding peaceful nuclear military activity. Thus, some critics may state that a Paragraph 14 arrangement does not require the board's approval. In this scenario, AUKUS partners would not need the approval of the board to engage in peaceful nuclear activity and its associated functions and may therefore not invoke Paragraph 14 exemptions. This means that none of the nuclear material used in Australia's submarine cores would be subject to safeguards.

It is important to note that each of these courses of action are largely speculative and could largely differ based on the findings presented at the culmination of the 18-month consultation period. However, the AUKUS partners would likely find great success in positively engaging with the IAEA to pursue a Paragraph 14 exemption outlined in Option 1 and receive a sealed reactor core from its security partners.

LEGAL RAMIFICATIONS FOR PEACEFUL NUCLEAR MILITARY ACTIVITY AND NUCLEAR TECHNOLOGY SHARING

Because the specific details guiding the partnership and information sharing entailed in AUKUS have not yet been determined or released to the public, it is difficult to determine where legal precedent stands to guide the AUKUS proposal to the IAEA.

The U.S. legal framework explicitly dictates the type of materials and information that can be shared or disclosed, the amount of time for the agreement to be in effect, and the ramifications for any abrogation of the treaty or pact in question. Regarding U.S. legislation, the Atomic Energy Act of 1954 regulates and outlines the permissible export of nuclear technology, materials, and expertise from the United States and will likely shape much of U.S. involvement in the AUKUS pact. The United States and United Kingdom already share a framework that regulates import and export activities and maintains a robust infrastructure to conduct information sharing between both countries via mutual defense agreements.[41] Section 54 of the Atomic Energy Act (AEA) authorizes the Atomic Energy Commission to "cooperate with any nation or group of nations by distributing special nuclear material ... pursuant to the terms of an agreement."[42] The AEA also authorizes the transfer of source material to foreign partners but limits the transfer to three metric tons per year.[43] This could support the logic of AUKUS pursuing Option 1 since the limitation of three metric tons would significantly stymie progress in nuclear reactor construction in Australia. Further, section 144 c.(2) of the AEA also stipulates several criteria that AUKUS partners must agree to in order to establish a legal nuclear cooperation agreement.[44] These stipulations include:

---

40   Rockwood, "Naval Nuclear Propulsion and IAEA Safeguards."

41   Office of Defense Programs, "Atomic Energy Act Control of Import and Export Activities," National Nuclear Security Administration, February 2015, https://www.energy.gov/sites/prod/files/migrated/nnsa/2017/11/f45/NAP-23%20AC1%20 Final%202-9-15.pdf.

42   Ibid; and Office of the General Counsel, *Nuclear Regulatory Legislation: 113th Congress; 2nd Session* (Rockville, MD: United States Nuclear Regulatory Commission, December 2015), https://www.nrc.gov/docs/ML1536/ML15364A497.pdf.

43   Ibid.

44   Ibid.

- a guarantee that safeguards on transferred nuclear materials and equipment continue in perpetuity;

- a provision requiring the application of comprehensive IAEA safeguards in the NNWS;

- a prohibition on the retransfer of material or restricted data without U.S. consent;

- a requirement that the recipient state maintain physical security of transferred nuclear material;

- a prohibition on the recipient state's use of transferred items or technology for any nuclear explosive device or for any other military purpose; and

- a provision specifying the United States' right to demand the return of transferred nuclear materials and equipment if the cooperating state detonates a nuclear explosive device or terminates an IAEA safeguards agreement.[45]

These stipulations could easily be applied to Option 1. Paragraph 14 still allows limited safeguards to the extent possible to protect sensitive reactor core and submarine designs. Further, Australia's credibility as a nonproliferation partner and NNWS suggests that the country is willing and able to abide by demands to apply any safeguards domestically if pursuing Option 1. Lastly, Australia is unlikely to pursue the illegal export of nuclear reactor material to an NWS or terminate an IAEA agreement. Should any of these extremes occur, the United States retains the right to terminate their role in the AUKUS agreement and confiscate all special nuclear material and reactor technology from Australia.

The legal ramifications regarding information sharing and data protection also leave little room for malicious or unauthorized access or disclosure. According to the Congressional Research Service (CRS), no party to the AUKUS defense pact may communicate any information to unauthorized persons beyond the party's jurisdiction or control.[46] Further, the CRS states that "a recipient party communicating such information to nationals of a third AUKUS government must obtain permission from the originating party." The report also includes an appendix which details the security arrangements set forth to ensure that transferred information is protected at all stages. Should any party choose to terminate its participation in the agreement or materially violate the agreement, the other AUKUS partners may "require the return or destruction" of any transferred data.[47] Additionally, the Exchange of Naval Nuclear Propulsion Information Agreement entered into force on February 8, 2022, enabling AUKUS partners to share naval nuclear propulsion information trilaterally.[48] These processes have enabled the safe and efficient transfer of data while ensuring that no AUKUS partner provides a third party with information that could support proliferation-permissive environments.

Legislation in the United Kingdom that may pertain to the AUKUS agreement is largely found in the U.S.-UK Mutual Defence Agreement (U.S.-UK MDA), which dictates much of the legal relationship both nations have regarding nuclear-propelled submarines and nuclear information sharing. Laws in the United Kingdom dictate that classified information received from partners will be protected according to specific arrangements outlined between the United Kingdom and the other nation in question.[49] The United Kingdom's Nuclear Industries Security Regulations 2003, Energy Act 2013, and

---

45   Paul Kerr and Mary Beth Nikitin, "AUKUS Nuclear Cooperation," Congressional Research Service, March 11, 2022, https://crsreports.congress.gov/product/pdf/IF/IF11999.

46   Ibid.

47   Ibid.

48   "Fact Sheet: Implementation of the Australia – United Kingdom – United States Partnership (AUKUS)," The White House, April 5, 2022, https://www.whitehouse.gov/briefing-room/statements-releases/2022/04/05/fact-sheet-implementation-of-the-australia-united-kingdom-united-states-partnership-aukus/.

49   "The Nuclear Industries Security Regulations 2003," UK Government, https://www.legislation.gov.uk/uksi/2003/403/

Atomic Energy Act of 1946 all provide guidance on nuclear technology and information sharing.[50] The United States and the United Kingdom may rely on pre-existing frameworks from the U.S.-UK MDA to govern much of the practice and implementation needed regarding Australia's submarine construction, delivery, and other cooperative measures. A similar framework, in conjunction with standing U.S. law, could be adapted for the purposes of the trilateral security agreement.

AMBIGUOUS RESPONSIBILITY

It can be inferred that there are various legal and technical loopholes that actors could potentially use to proliferate nuclear material for uses other than peaceful nuclear military activity. A framing question that AUKUS stakeholders must be vigilant about is if other IAEA member states would be comfortable with the security pact's framing being used among other varieties of NWSs and NNWSs. N^{th}-country portfolios that could feasibly handle the technical and infrastructure requirements of acquiring nuclear-powered submarines are dutifully taking notes and keeping track of the progress made by AUKUS partners during the 18-month consultation period. Whatever precedent is set by the security pact must uphold the ethical, technical, political, and military norms set by the rules-based order and the NPT, regardless of which countries are involved.

It may be too early to tell, however, which provisions will be most relevant until the specific details of AUKUS are worked out during the ongoing 18-month coordination period. Unknown factors at this stage include how the United States and United Kingdom plan to delineate responsibility for the supply of nuclear materials to Australia; the mechanisms for the transfer, storage, and training associated with such materials; fuel fabrication and export schemes; the intended enrichment process (as applicable); and the accompanying IAEA safeguards that will be required to ensure the safety and security of the materials. The largest loophole still remains as to when and if Australia's nuclear submarine cores will be subject to safeguards from the IAEA. If following the precedent set by the U.S. and UK nuclear submarine programs, the nature of the classification of the submarines could require the cores be exempt from safeguards for stretches at a time. Because n^{th}-country portfolios could seek to exploit this loophole in the IAEA's definition and requirements of HEU material used for naval nuclear propulsion, skeptics still see AUKUS's invocation of Paragraph 14 as not sufficient enough to uphold nonproliferation norms. Unless the AUKUS pact voluntarily subjects itself to safeguards or relies on a large amount of additional protocols, CSAs, and other ancillary documents to provide information that would quell proliferation concerns, AUKUS's intent to deliver nuclear-powered submarines will likely still be seen as highly controversial and indecorous. Given that all three nations are firm supporters of the NPT, the rules-based order, and nonproliferation norms, AUKUS stakeholders must wrestle with the reality of other n^{th}-country portfolios emulating their program.

Without credible answers to these questions, much of the information regarding the AUKUS pact should be recognized as inferences. Thus, it is difficult to determine which governing body, state, international organization, or private entity should handle or facilitate various aspects of the security agreement.

According to INFCIRC/153, however, the IAEA is responsible for providing approval as soon as possible—and for only providing agreement regarding the matters which have been presented by

part/1/made.

50   Ibid.; "Energy Act 2013," UK Government, December 2013, https://www.legislation.gov.uk/ukpga/2013/32/contents/enacted/data.htm; and "Atomic Energy Act 1946," UK Government, March 2021, https://www.legislation.gov.uk/ukpga/Geo6/9-10/80/data.pdf.

states and not on matters regarding classified or sensitive material.[51] To do so, the Board of Governors must agree on all aspects of the trilateral defense pact's mission to deliver submarines to Australia. The board can either vote publicly or through a secret ballot and can request that proposals or amendments be voted on separately. Voting inclinations could be skewed by the voting rights of an NWS who possesses naval nuclear technology; states such as China may feel that their status as an NWS is threatened by the successful construction and delivery of submarines to Australia and may be inclined to vote against the proposal and encourage others to do the same. Other member states, if skeptical about the IAEA's approval process, may view the invocation of Paragraph 14 as a nicety rather than a required part of the dialogue between the agency and the state. The IAEA will likely need to take significant steps to ensure that the Board of Governors and Secretariat are well informed of key details regarding the proposal while still maintaining the necessary information requirements regarding security and sensitive details. The trilateral security pact should also seek the feedback of technical subject matter experts and other stakeholders to ensure that the review process remains holistic and does not undermine the integrity of the IAEA.

The IAEA must bear the responsibility of ensuring that a sufficient amount of detail is provided so that the Board of Governors can make a decision, to the best of their ability, without bias or outside influence. Since the AUKUS deal will set a new precedent regarding nuclear propulsion technology sharing and the export of nuclear material to an NNWS, the IAEA must take all steps to ensure that the process maintains its integrity so that current international norms are not undermined throughout the process.

CONCLUDING THOUGHTS

FOMENTING DISTRUST AND OTHER GEOPOLITICAL TENSIONS

Though Australian defense minister Peter Dutton stressed that "Australia is not seeking nuclear weapons," many countries have posed concerns regarding the export of special nuclear material and the potential threats to international nonproliferation efforts.[52] In a memo to the IAEA, China requested a special meeting of IAEA members to discuss a potential safeguards regime for naval nuclear technology; the country also expressed a desire for "AUKUS partners [to] refrain from commencing their cooperation until a safeguards system is in place."[53] Chinese Foreign Ministry spokesman Zhao Lijian also described the AUKUS defense pact as "extremely irresponsible . . . [to] advance nuclear submarine cooperation in disregard of international rules and opposition of parties."[54] NWSs not aligned with the AUKUS defense pact may view the submarine acquisition aspect of the agreement as one of the most harmful facets for stability and security in the Indo-Pacific region. Careful consideration should be given to the importance of varying perceptions regarding naval nuclear propulsion; the trilateral partnership will likely need to continue to engage with the IAEA director and Board of Governors through bilateral, trilateral, and even multilateral fora to establish trust and understanding throughout the 18-month period. These efforts should also include meetings to raise the awareness of other IAEA member states and regional partners regarding the impact AUKUS intends to have on the naval nuclear submarine regime. Simultaneously, AUKUS signatories should move to bolster a public information campaign to explain the benefits of the submarine acquisition and address concerns that exist among nuclear critics and the public. The

---

51 "INFCIRC/153(Corrected)," IAEA.
52 Julia Masterson, "AUKUS States Sign Information Exchange Deal," Arms Control Today, 2022, https://www.armscontrol.org/act/2022-01/news/aukus-states-sign-information-exchange-deal.
53 Ibid.
54 Ibid.

trilateral partners can utilize White House fact sheets, press releases, and other media to continue to inform and publicize certain aspects of the agreement to encourage discussion and show progress in their determination to enact the submarine acquisition portion of the deal. A concerted effort to focus on the positive impacts on defense, technology-sharing precedents, cooperation, stability, and economic growth could shift negative perceptions regarding the AUKUS agreement.

The AUKUS security pact also has implications for the geopolitical environment within the Indo-Pacific. According to Hans Kristensen, director of the Nuclear Information Project, the AUKUS deal will "intensify the arms race in the region and [exacerbate] dynamics that fuel military competition."[55] Any increase in military equipment or strategic force in the Indo-Pacific will likely be perceived as an aggressive and offensive posture by Chinese leaders. Additionally, regimes with a hardline stance toward the West, such as Iran, will continue to view the AUKUS defense pact as a double standard benefitting countries within the "Anglosphere," or Five Eyes regime, an intelligence alliance comprised of Australia, Canada, New Zealand, the United States, and the United Kingdom.[56]

The AUKUS pact may also further exacerbate the asymmetric nature that exists within the NPT between NWSs and NNWSs. Though the NPT does not prohibit NNWSs from pursuing nuclear-powered naval technology, many of the states in the Indo-Pacific region do not possess nuclear weapons or nuclear weapons technology.[57] Australia's new nuclear submarine fleet could act as an extension of the "strategic diplomacy umbrella" in protecting assets ashore and at sea via conventional weapons launched by nuclear submarines. Vulnerable partner nations who may be at risk of debt-trap diplomacy or aggressive financial or military pressure from China may be assured of a robust Australian presence in the region as a natural extension of defense. Moreover, the continually changing security environment forces states such as Australia to pursue stabilizing defense postures that include the acquisition of modern technology that can respond to the nature of pacing challenges and acute threats.

PRECEDENT IN EXCHANGE FOR INTERNATIONAL NORMS

A successful proposal by the trilateral partnership must limit the potential for any nuclear material to be lost to any nonstate actors, illicit state intentions, or nonprescribed peaceful military means. Additionally, the arrangement should clearly delineate when states intend to invoke the nonapplication of safeguards for the use of propulsion and when safeguards will be re-invoked in line with reporting and record-keeping criteria. Finally, the agreement should propose the use of a combination of Subsidiary Arrangements and ancillary documents to disclose sensitive details to the IAEA, as required, while maintaining the safety and integrity of the naval nuclear propulsion program. Any arrangement that allows the withdrawal of special nuclear material necessarily will be confidential in its nature.

Apart from Canada's initial approach in the late 1980s to conclude such an arrangement with the IAEA, which Canada terminated in 1990, the IAEA has not been called upon to address the legal and technical aspects of such arrangements. The AUKUS tripartite agreement would therefore set an important precedent. Brazil's pledge to commission a nuclear-powered submarine by 2029 is also relevant but lacks the defense pact's commitment to data and technology sharing; a lack of support, technical difficulties, and various obstacles have slowed the nuclear innovation process within Brazil.

55 Anastasia Kapetas, "Editors' Picks for 2021: 'Limiting the Nuclear-Proliferation Blowback from the AUKUS Submarine Deal,'" Australian Strategic Policy Institute, The Strategist, January 4, 2022, https://www.aspistrategist.org.au/editors-picks-for-2021-limiting-the-nuclear-proliferation-blowback-from-the-aukus-submarine-deal/.

56 J. Vitor Tossini, "The Five Eyes - the Intelligence Alliance of the Anglosphere," UK Defence Journal, April 14, 2020, https://ukdefencejournal.org.uk/the-five-eyes-the-intelligence-alliance-of-the-anglosphere/.

57 Excluding North Korea, China, India, and Pakistan.

Both the Canadian and Australian proposals open a realm of possibilities for utilizing the cut-off of safeguards in Paragraph 14 of INFCIRC/153 for future innovation and discovery. Under this framework, countries could theoretically pursue reactors for military research or design propulsion rockets for vehicles built for space flight. However, the IAEA may see further utilization of Paragraph 14 as a vehicle to pursuing proliferation regimes which attack the international nonproliferation regime and allow n^{th}-country portfolios to continue to pursue advanced proliferation technology and nuclear weapon capabilities. Because of Australia's track record, it is unlikely that the IAEA, or the international community of NPT signatories, will be alarmed by Australia's pursuit of a nuclear submarine fleet for the same concerns. Moreover, from a legal standpoint, the IAEA would likely not have much dearth to reject the AUKUS proposal to nuclear material withdrawal.

The AUKUS defense pact's potential invocation of Paragraph 14 should not stand as precedent for nuclear fringe states to pursue volatile nuclear weapons programs. Instead, the security pact will set precedent for allies to share technology, materials, and resources in overcoming future challenges to maintaining global peace and stability. The AUKUS deal is a chance to exercise confidence-building measures among NWSs and NNWSs and facilitate a way to alleviate the asymmetrical relationship that exists between safeguards and NWS verification. Assuming the trilateral partnership will continue to acknowledge and abide by IAEA safeguards and conduct due diligence regarding special nuclear materials, the AUKUS pact should not be viewed as a step backwards in nonproliferation standards. Instead, the AUKUS pact champions efforts to positively engage countries with a strong reputation in nuclear safeguards compliance to join the nuclear-powered submarine regime. AUKUS sets a positive precedent that gives a fresh outlook to the future of nuclear nonproliferation discussions. The nuclear community must champion and encourage efforts to pursue safe, efficient, and innovative methods of naval nuclear propulsion to generate continual interest and vitality in nuclear technologies through the twenty-first century and beyond.

The U.S. Position in Global Nuclear Exports

By Caitlyn Bess[1]

INTRODUCTION

On May 22, 2020, Russia's first small modular reactors (SMRs) aboard a floating nuclear power plant became fully operational and began supplying power to the town of Pevek, Russia.[2] In December 2021, China announced that its first high-temperature gas-cooled reactor (HTGR) featuring a pebble-bed module had been connected to the electricity grid, becoming the world's first commercial HTGR of the pebble bed design (HTR-PM).[3] Also in December 2021, the United States' leading SMR company announced in a press release that it is positioned to deliver its first power plant to a customer, but in 2027 at the earliest.[4]

The dates in these news stories illustrate the larger trend that the United States has fallen behind Russia and China in deploying new forms of nuclear energy. While the United States is still home to a world-class nuclear energy industry in terms of innovation, that industry is behind when it comes to building reactors—whether traditional large light-water reactors (LWRs) or latest-generation advanced and small modular reactors (ASMRs).

---

1   Caitlyn Bess is a systems implementation and integration professional at Sandia National Laboratories (SNL), where she works with international partners cultivating and maintaining strong nuclear security cultures. The views expressed in this paper are her own and do not necessarily reflect the views of her employer. The author thanks her colleagues Shannon Abbott and Jennifer Obrey-Espinoza and Andrea Viski for their comments and edits on this paper.

2   "ROSATOM: world's only floating nuclear power plant enters full commercial exploitation," Rosatom, May 22, 2020, https://rosatom.ru/en/press-centre/news/rosatom-world-s-only-floating-nuclear-power-plant-enters-full-commercial-exploitation/.

3   "Demonstration HTR-PM connected to grid," World Nuclear News, December 21, 2021, https://www.world-nuclear-news.org/Articles/Demonstration-HTR-PM-connected-to-grid.

4   "Press Release: NuScale Power, the Industry-Leading Provider of Transformational Small Modular Nuclear Reactor Technology, Announces Plans to Go Public via Merger with Spring Valley Acquisition Corp," NuScale, Dec 14, 2021, https://newsroom.nuscalepower.com/press-releases/news-details/2021/NuScale-Power-the-Industry-Leading-Provider-ofTransformational-Small-Modular-Nuclear-Reactor-Technology-Announces-Plans-to-Go-Public-via-Merger-with-Spring-Valley-Acquisition-Corp/default.aspx.

This paper will first demonstrate that the United States is behind both Russia and China in this industry, in both LWR and ASMR deployment. Second, it will explore three areas where this dynamic is concerning for the United States: (1) the growing strategic importance of nuclear energy, (2) the strengthening of Russia and China's nuclear exports, and (3) the increasing ability of both countries to influence the nuclear safety and security cultures in the countries where they export. Finally, the paper will account for recent U.S. government efforts to bolster research and development (R&D) and exports in this industry and then end with policy recommendations to enhance those efforts.

THE STATE OF THE INDUSTRY

To demonstrate that the U.S. nuclear energy industry is behind the state-owned Russian and Chinese industries, it is useful to examine the Power Reactor Information System (PRIS) published by the International Atomic Energy Agency (IAEA). Every power reactor constructed or under construction is documented on the PRIS, including information about each reactor's design. This design can be traced back to the nationality of the company where it originated, and from there one can compare how many reactors worldwide are of U.S., Russian, or Chinese design. These figures can then serve as a quantitative proxy for the influence each country has in the global nuclear energy industry.

The number of reactors of U.S., Russian, and Chinese design operating in third countries is listed in Table 1. Since this paper is primarily interested in nuclear energy exports and the strategic advantage that comes with them, reactors in the United States, China, and Russia were excluded.

Table 1: Reactor Exports Worldwide (excluding to the United States, China, and Russia)

Vendor Country	Operational	Under Construction	Connected to Grid Post-2000
United States	90	2*	0
Russia	38	16	8
China	6	2**	6
Other Vendors***	104	13	27
Total Reactors, excluding to the United States, China, and Russia	238	33	41

Source: "Power Reactor Information System (PRIS)," IAEA, 2022, https://www.iaea.org/resources/databases/power-reactor-information-system-pris.

* The two reactors currently under construction are the ABWR model, a design from GE-Hitachi, which is a U.S.-Japanese partnership.

** These reactors are Hinkley Points C-1 and C-2 in the United Kingdom. Their design is not Chinese, but the state-owned energy corporation China General Nuclear (CGN) is one of the plant's major investors. At the time of this writing, whether CGN will remain in the deal is uncertain.5

***This category includes the exports from countries other than the United States, Russia, and China and domestic nuclear reactor construction by countries other than the United States, Russia, and China.

This table tells a story of a global industry that has waned in the past few decades, as only a small percentage of total operational reactors have been connected to the grid post-2000. However, of the

5   Ashish Dangal, "After £20B 'British Snub' To China, Another Nuclear Project Faces The Heat Over Chinese Involvement," *Eurasian Times*, May 10, 2022, https://eurasiantimes.com/china-nuclear-project-faces-the-heat-over-chinese-involvement/.

reactors that are currently under construction or connected to the grid since 2000, U.S. companies supplied none that are still in operation in 2022, and Russia and China supplied an increasing proportion, especially Russia.

All reactors counted in Table 1 are traditional large LWRs, but it is also important to examine the ASMR landscape. ASMRs are likely to constitute a large part of the nuclear energy market in coming decades because of the interest many countries have shown in them, so suppliers which can bring these latest-generation designs to market the quickest will be at an advantage in the export market. imilar to the PRIS, the IAEA's Advanced Reactor Information System (ARIS) provides information to assess the global ASMR landscape. Table 2 updates data originally from 2020 from ARIS and shows the progress toward deployment of the seven most developed ASMR projects worldwide.

Table 2: Highly Developed ASMR Projects Worldwide

Country	Reactor	2019	2020	2021	2022	2023	2024	2025	2026	2027	2028	2029
Russia	KLT-40S	Commercial operation										
China	HTR-PM	Construction		Demonstration operation								
Argentina	CAREM25	Construction				Prototype operation						
China	ACP100	Licensing		Construction				Prototype operation				
United States	NuScale	NRC approval	limbo			Production and construction				Commercial operation		
Russia	RITM-200M	Licensing					Construction				Operation	
South Korea	SMART	Licensing					Construction				Operation	

Source: Data from "Figure I-2 General Timeline of Deployment as of 2020," from IAEA, *Advances in Small Modular Reactor Technology Developments* (Vienna: 2020), 308, https://aris.iaea.org/Publications/SMR_Book_2020.pdf. Updates from "NRC Interaction," NuScale, July 1, 2021, https://www.nuscalepower.com/technology/licensing; "Full-scale production of NuScale SMR to begin," World Nuclear News, April 25, 2022, https://www.world-nuclear-news.org/Articles/Full-scale-production-of-NuScale-SMR-to-begin; "Press Release: NuScale and Nuclearelectrica Reach Agreement at COP26 to Initiate the Deployment of the First Small Modular Reactor in Europe," NuScale, November 4, 2021, https://newsroom.nuscalepower.com/press-releases/news-details/2021/NuScale-and-Nuclearelectrica-Reach-Agreement-at-COP26-to-Initiate-the-Deployment-of-the-First-Small-Modular-Reactor-in-Europe/default.aspx; "Demonstration HTR-PM connected to grid," World Nuclear News, December 21, 2021, https://www.world-nuclear-news.org/Articles/Demonstration-HTR-PM-connected-to-grid; and "China starts construction of demonstration SMR," World Nuclear News, July 13, 2021, https://world-nuclear-news.org/Articles/China-starts-construction-of-demonstration-SMR.

This table shows that, as of this writing, there are only two currently operating ASMRs in the world—one in Russia and one in China. The KLT-40S reactors are on board the *Akademik Lomonosov*, Russia's floating nuclear power plant in Pevek, Russia. The HTR-PM was connected to the grid in December 2021 at the Shidaowan power plant in China. Argentina has been constructing its prototype CAREM25 reactor since 2014, and it is projected to enter operation in 2023. China started construction on its ACP100 reactors in July 2021. The only U.S. company to make the top seven, NuScale, earned Nuclear Regulatory Commission (NRC) approval of its design in August 2020, 12 years after the first notification to the NRC that it would pursue design certification.[6] In April 2022, it signed an agreement with Doosan Enerbility to begin manufacturing main equipment which will eventually go into its first projects, estimated to begin operation in 2027 at the earliest.[7]

6   "History of NuScale Power," NuScale, https://www.nuscalepower.com/about-us/history.
7   "Full-scale production of NuScale SMR to begin," World Nuclear News, April 25, 2022, https://www.world-nuclear-news.

Tables 1 and 2 both indicate that Russia and China are in a better position than the United States to take advantage of growing interest in nuclear energy in many parts of the world. Both Russia and China have more recent experience with nuclear energy projects at home and abroad, helping to maintain relevant skills and supply chains for these multi-billion-dollar projects. Operating ASMR plants also puts them at an advantage, as operating experience is a strong selling point for first-of-a-kind designs. While there is immense interest in ASMRs, few countries want to build them first.

A STRATEGIC LIABILITY

The fact that the United States is behind its largest geopolitical competitors in this industry is concerning for three main reasons. First, nuclear energy will likely be a strategically important global industry in the future. Second, Russia and China are economically and geopolitically strengthened by their nuclear energy exports. Third, Russia and China will influence the nuclear safety and security cultures that take root in the countries they export to, especially if they are nuclear newcomer countries. The following sections will explore each of these issues more deeply.

A STRATEGIC GLOBAL INDUSTRY

The United States falling behind in nuclear energy exports is concerning because it may lose its ability to compete in a strategic global industry. The think tank Third Way and the Energy for Growth Hub estimate that the global market for nuclear power could triple by 2050.[8] According to the consulting group Spherical Insights, the market was estimated at about $33 billion in 2021, so by 2050 it could be a $100 billion industry.[9] Third Way and the Energy for Growth Hub predict most of that growth will be outside of high-income countries and in nuclear newcomer countries. The IAEA writes on the ARIS homepage that "the global demand for electricity will continue to increase as a result of economic development aspirations and energy security concerns, and that expansion of nuclear power will continue at a slow but steady pace particularly in Asia and the Middle East."[10] Because energy is a strategic asset no matter how it is generated, it is in the United States' interest to be a supplier of that energy.

Considering the many reasons countries are interested in nuclear energy programs, this projected growth suggests that nuclear energy will help countries achieve their carbon-reduction goals and enhance energy security. The impetus to find carbon-emission-free sources of reliable baseload electricity is more urgent than ever as climate change, driven by greenhouse gases, affects more and more people.[11] However, popular forms of carbon-free electricity generation such as photovoltaic solar panels and wind turbines do not provide baseload electricity. Nuclear energy is a proven source of carbon-free baseload electricity, so many countries are making concrete steps toward reinvesting in or starting new nuclear energy programs to reduce their carbon emissions.[12] According to the IAEA, around 30 countries were considering or embarking on nuclear energy programs as of November

---

org/Articles/Full-scale-production-of-NuScale-SMR-to-begin.

8   "2021 Update: Map of the Global Market for Advanced Nuclear," Third Way, November 9, 2021, https://www.thirdway.org/memo/2021-update-map-of-the-global-market-for-advanced-nuclear.

9   "Global Nuclear Power Plant Equipment Market Size to grow USD 38.82 Billion by 2030 | CAGR of 2.6%," Spherical Insights LLP, October 2, 2022, https://www.globenewswire.com/news-release/2022/10/02/2526412/0/en/Global-Nuclear-Power-Plant-Equipment-Market-Size-to-grow-USD-38-82-Billion-by-2030-CAGR-of-2-6.html.

10   "Future of Nuclear Power," International Atomic Energy Agency, https://aris.iaea.org/.

11   Intergovernmental Panel on Climate Change, *Climate Change 2022: Impacts, Adaptation and Vulnerability: Summary for Policymakers* (Cambridge, UK: Cambridge University Press, 2022), 8–12, 33, https://www.ipcc.ch/report/ar6/wg2/.

12   While there are carbon emissions associated with the construction of nuclear plants (as there would be in the construction of any type of facility), they do not produce carbon during operation.

2020, and as of this writing, of the 18 countries worldwide in which reactors are under construction, five of them are building reactors for their first-ever nuclear power plants.[13] ASMRs have caught even more countries' attention because of their smaller capital costs and energy output, thus making them potentially accessible to countries which could not build traditional large nuclear plants. In recent years, the United States has been engaging with nuclear newcomers from all over the world, specifically toward the goal of ASMR deployment.[14]

Countries are also interested in nuclear energy for energy security. Energy is a critical industry for any country because electricity enables a modern economy and supports the health and well-being of citizens. Russia's renewed invasion of Ukraine in 2022 and Russia's weaponization of natural gas exports has served as a reminder to governments around the world of the strategic vulnerability of relying on another state, and particularly a hostile state, to supply electricity. Nuclear energy increases energy security because uranium is an extraordinarily energy-dense resource. One kilogram of mined natural uranium will yield about 20,000 times as much energy as one kilogram of mined coal.[15] This means reactors only need to be refueled every few years rather than daily, thus making nuclear-generated electricity less dependent on geopolitical conditions.

While low-carbon electricity generation and energy security are the primary reasons countries are interested in nuclear energy, there are more advantages to nuclear energy related to ASMRs in particular. First, they are cheaper than traditional nuclear power plants because of their smaller size. Their modularity should also shorten construction timelines, which makes the financing costs lower.[16] ASMRs are also particularly well suited to developing countries whose electricity grids cannot immediately absorb the power output of a traditional large nuclear power plant. ASMR plants also offer more siting flexibility. They could be installed in former coal plants, which already have the necessary grid connections, or sited in remote regions to offer power to isolated communities, mines, or other industrial projects.[17] The process heat from ASMR plants could also be harnessed to power hydrogen production, desalination, or other processes that require extreme heat. ASMRs are also safer due to their smaller size, passive safety systems, and advanced coolants and fuel types.[18] These enhanced safety features additionally may lead to broader public acceptance of ASMR plants.

13   Matthew Fisher, "Developing Nuclear Power Infrastructure in Newcomer Countries," IAEA, *IAEA Bulletin* vol. 61, no. 4, November 2020, https://www.iaea.org/bulletin/developing-nuclear-power-infrastructure-in-newcomer-countries; and "Under Construction," IAEA, Power Reactor Information System (PRIS), https://pris.iaea.org/PRIS/WorldStatistics/UnderConstructionReactorsByCountry.aspx.

14   For examples of nuclear newcomer countries engaging with the United States on deploying SMRs, see "Estonia joins U.S. FIRST program," U.S. Embassy in Estonia, January 24, 2022, https://ee.usembassy.gov/2022-01-24/; "Joint Statement on the New Clean Energy and Nuclear Security Collaboration under the Foundational Infrastructure for Responsible Use of Small Modular Reactor Technology (FIRST) Initiative," U.S. Department of State, April 4, 2022, https://www.state.gov/joint-statement-on-the-new-clean-energy-and-nuclear-security-collaboration-under-the-foundational-infrastructure-for-responsible-use-of-small-modular-reactor-technology-first-initiative/; "U.S. and Ghana Take Next Steps Forward on Clean, Safe Nuclear Energy," U.S. Embassy in Ghana, March 8, 2022, https://gh.usembassy.gov/u-s-and-ghana-take-next-steps-forward-on-clean-safe-nuclear-energy/; and "The United States Announces $25 Million to Support Access to Clean Nuclear Energy," U.S. Department of State, November 3, 2021, https://www.state.gov/the-united-states-announces-25-million-to-support-access-to-clean-nuclear-energy/.

15   "The Economics of Nuclear Power," World Nuclear Association, November 2008, 13, https://www.world-nuclear.org/uploadedfiles/org/info/pdf/economicsnp.pdf.

16   IAEA, *Technology Roadmap for Small Modular Reactor Deployment* (Vienna: August 2021), IAEA Nuclear Energy Series No. NR-T-1.18, https://www.iaea.org/publications/14861/technology-roadmap-for-small-modular-reactor-deployment.

17   Nicholas Watson and Nikoleta Morelova, "Repurposing Fossil Fuel Power Plant Sites with SMRs to Ease Clean Energy Transition," IAEA, June 16, 2022, https://www.iaea.org/newscenter/news/repurposing-fossil-fuel-power-plant-sites-with-smrs-to-ease-clean-energy-transition; and ibid., 19.

18   IAEA, Technology Roadmap for Small Modular Reactor Deployment.

The world could be on the brink of a second nuclear energy construction boom, but the United States is not currently poised to take advantage of it. The demand for nuclear energy is growing worldwide such that it will likely constitute a significant percentage of the global energy mix in the future, and those nations which become major exporters will reap the geopolitical advantages that naturally come with being an energy supplier.

CHINESE AND RUSSIAN STRENGTH IN NUCLEAR EXPORTS

The Russian and Chinese governments seem to recognize the financial and strategic advantages of supplying other countries with nuclear power plants. First, Russia uses its nuclear reactor and fuel exports to generate revenue for the Russian state. Rosatom is the world's biggest exporter of nuclear reactors. Its revenues from exports doubled in the past 10 years and make up 40 percent of its total revenue.[19] It claims that its package of foreign orders exceeded $140 billion dollars in 2021, suggesting a steady flow of future revenue.[20] Russia also dominates global nuclear fuel production. Columbia University's Center on Global Energy Policy released a report in May 2022 which found that Russia is a major global player at every stage of nuclear fuel production and the dominant one in the conversion and enrichment steps.[21] It would take many years for other countries to fill Russia's void if it were to weaponize nuclear fuel the same way it has weaponized natural gas.[22] Russia tends to insert exploitative terms into its international dealings, such as gaining control over the natural resources of "customer" countries.[23] Russia also seems to expect political support from customer countries and political leverage over them. For example, its customer countries in Asia, South America, and Africa in 2022 resisted U.S. efforts to rally them behind economic sanctions on Russia and military support for Ukraine; fear of being cut off from Russian exports was likely a part of that reticence.[24] The March 2022 UN resolution condemning Russia's actions in Ukraine was touted for 141 of 193 nations voting for it, but 35 countries abstained, and among those 35 are customers of Russian commodities.[25] It is reasonable to wonder if countries which are dependent on Russia in any way, including nuclear energy, would hesitate to cross it, given that it has proven via its natural gas dealings with Europe in 2022 that it is willing to weaponize exports. It is perhaps for these reasons a 2019 document by the Stockholm International Peace Research Institute (SIPRI) called Russia's exported nuclear power plants a "strategic asset" and a director at the Nuclear Energy Institute in Washington called Rosatom a "designated strategic exporter."[26]

19   "Rosatom Highlights in 2021," Rosatom, January 31, 2022, https://rosatomnewsletter.com/2022/01/31/rosatom-high-lights-in-2021/.

20   "Projects," Rosatom, n.d., https://rosatom.ru/en/investors/projects/.

21   Matt Bowen and Paul Dabbar, "Reducing Russian Involvement in Western Nuclear Power Markets," Columbia University, SIPA Center on Global Energy Policy, May 23, 2022, https://www.energypolicy.columbia.edu/research/commentary/reduc-ing-russian-involvement-western-nuclear-power-markets.

22   Ibid.

23   Brian Katz et al., "Moscow's Mercenary Wars: The Expansion of Russian Private Military Companies," CSIS, September 2020, https://russianpmcs.csis.org/.

24   Lara Jakes and Edward Wong, "Biden Races to Expand Coalition Against Russia but Meets Resistance," *New York Times*, June 11, 2022, https://www.nytimes.com/2022/06/11/us/politics/russia-biden-sanctions.html.

25   "Aggression against Ukraine: Voting Summary," UN General Assembly, March 2, 2022, https://digitallibrary.un.org/re-cord/3959039.

26   Nevine Schepers, "Russia's Nuclear Energy Exports: Status, Prospects and Implications," Stockholm International Peace Research Institute, Non-Proliferation and Disarmament Papers No. 61, February 2019, https://www.sipri.org/sites/default/files/2019-02/eunpdc_no_61_final.pdf; and Ivan Nechepurenki and Andrew Higgins, "Coming to a Country Near You: A Russian Nuclear Power Plant," *New York Times*, June 22, 2020, https://www.nytimes.com/2020/03/21/world/europe/belar-us-russia-nuclear.html.

China is not as likely as Russia to use nuclear exports to generate state revenue, but it is similar to Russia in the expectations it puts on its "partner" nations which are recipients of its loans, contracts, and aid. For example, China has a pattern of offering infrastructure projects in exchange for other countries breaking diplomatic ties with Taiwan, which has largely been successful.[27] It also offered the island nation of Tuvalu loans and assistance for building artificial islands, which Tuvalu rejected (for now) partly out of concern China would demand to establish military bases.[28] Participants in China's Belt and Road Initiative (BRI) are often under similar pressure.[29] Perhaps the most well-known example of an exploitative BRI infrastructure project is the Hambantota port in Sri Lanka, which is now de facto Chinese and which it has used for military purposes.[30] There is no reason to expect China would adopt a different modus operandi for its nuclear energy exports. While China has only ever constructed six nuclear reactors abroad (all in Pakistan, home to another Chinese port), there is evidence the Chinese government anticipates nuclear power being a more prominent export in the future.[31] China is touting its homegrown Hualong-1 reactors specifically for export. Two of the six reactors at the Karachi nuclear power plant in Pakistan are of that design, and Argentina could be the second country outside of China to commit to the Hualong-1, with construction set to start on the Atucha III site in 2022.[32] China's state-run nuclear energy company said of the Hualong-1, a traditional reactor, "Qualified for export, it has lifted China's nuclear power technology to a globally leading position and become a strategic tool for the government to promote nuclear power technologies abroad."[33] It is too early to tell whether the Hualong-1 will become the major export product the Chinese government would like it to be, but that is the goal.

Both China and Russia do not have qualms about using international infrastructure projects to their own advantage. Although the nuclear energy contracts are confidential, there is no reason to expect China and Russia would consider these different from other projects. Their nuclear dealings are numerous and widespread, as Figure 1 shows.

---

27   Thomas J. Shattuck, "The Race to Zero?: China's Poaching of Taiwan's Diplomatic Allies," *Orbis* 64, no. 2 (March 2020): 352, doi:10.1016/j.orbis.2020.02.003.

28   Ibid.

29   For China's dealings with BRI partner nations, see Paul Nantulya, "Implications for Africa from China's One Belt One Road Strategy," Africa Center for Strategic Studies, March 22, 2019, https://africacenter.org/spotlight/implications-for-africa-china-one-belt-one-road-strategy/; Jacob J. Lew et al., *China's Belt and Road: Implications for the United States* (New York, NY: Council on Foreign Relations, 2021), https://www.cfr.org/report/chinas-belt-and-road-implications-for-the-united-states/download/pdf/2021-03/TFR_Belt%20and%20Road%20Initiative_03172021_SinglePages.pdf.

30   Don M. Gill, "Geopositional Balancing: Understanding China's Investments in Sri Lanka," *Journal of Indo-Pacific Affairs* (Fall 2021): 109–13, https://media.defense.gov/2021/Aug/24/2002838283/-1/-1/1/GILL.PDF; and Uditha Jayasinghe, "Chinese military ship docks at Sri Lanka port despite Indian concern," Reuters, August 16, 2022, https://www.reuters.com/world/asia-pacific/chinese-military-survey-ship-docks-sri-lanka-port-2022-08-16/.

31   Gurmeet Kanwal, "Pakistan's Gwadar Port: A New Naval Base in China's String of Pearls in the Indo-Pacific," CSIS, *CSIS Briefs*, April 2, 2018, https://www.csis.org/analysis/pakistans-gwadar-port-new-naval-base-chinas-string-pearls-indo-pacific.

32   "China inks $8 bln nuclear power plant deal in Argentina," Reuters, February 2, 2022, https://www.reuters.com/business/energy/china-inks-nuclear-power-plant-deal-with-argentina-2022-02-02/.

33   "News," China National Nuclear Corporation, https://en.cnnc.com.cn/HPR1000.html.

Figure 1: Countries with Russian- and Chinese-Supported Nuclear Energy Projects

Source: Author's own creation. Data from "Russia and China Are Dominating Nuclear Energy Exports. Can the U.S. Catch Up?," Nuclear Energy Institute, 2021, https://www.nei.org/CorporateSite/media/filefolder/resources/fact-sheets/Russia-and-China-Are-Dominating-Nuclear-Energy-Exports-Can-the-U-S-Catch-Up.pdf.

China and Russia are both hostile to the United States, so it is not in the United States' interests to see them strengthened by nuclear energy exports or customer countries put in precarious situations through Chinese or Russian imports. One of the most effective ways the United States could undercut the strong Russian and Chinese positions in this market is to make it easier for potential customer countries to buy U.S. reactors.

SAFETY AND SECURITY NORMS

The final reason it is concerning that the United States is behind China and Russia in the global nuclear energy industry is that China and Russia will influence the nuclear safety and security norms in the countries to which they export. Nuclear safety culture and nuclear security culture cannot be sold or bought—they can only be built slowly over time. That is because they ultimately rest on a foundation of beliefs and attitudes, as shown in Figure 2, taken from the IAEA's Nuclear Security Series No. 7, *Nuclear Security Culture*. The relationships between a nuclear power plant supplier and owner are long term—from planning to construction to operation to decommissioning, a single power plant project could easily span 100 years or more. The multigenerational nature of this relationship gives ample time to impart beliefs and attitudes and to build (or not build) strong nuclear safety and security cultures.

Nuclear safety and security culture is an area in which the United States excels. The standards to which U.S. nuclear power plants are held are some of the strictest in the world. Public safety is a priority, and the belief that security threats to nuclear power plants are real is pervasive. This is probably why nuclear power plants in the United States are safer than other power generation methods in the country.[34]

---

34 See "U.S. Nuclear Industry Safety Accident Rate: One-Year Industry Values," Nuclear Energy Institute, December 2021, https://www.nei.org/resources/statistics/us-nuclear-industrial-safety-accident-rate; and David Brown, "Nuclear power is

Figure 2: Nuclear Security Culture

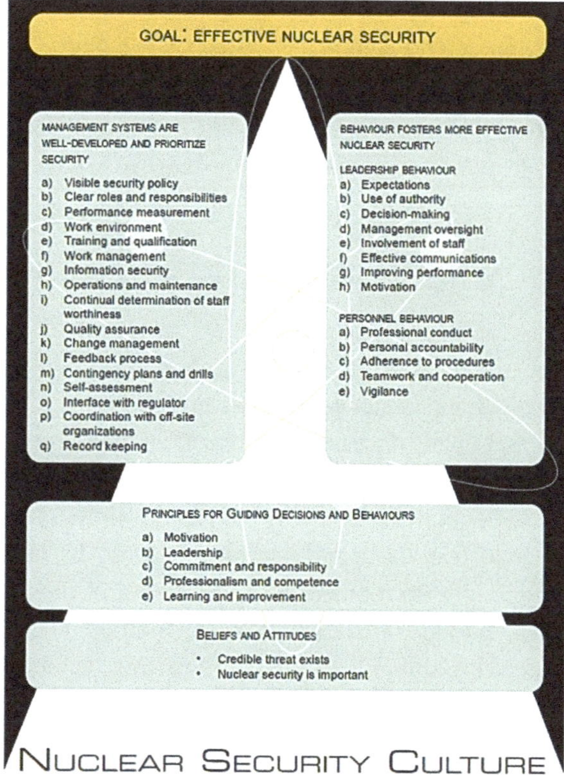

Source: © IAEA, *Nuclear Security Culture: Implementing Guide* (Vienna: 2008), IAEA Nuclear Security Series No. 7, 18, https://www-pub.iaea.org/MTCD/Publications/PDF/Pub1347_web.pdf.

There are reasons to suspect China and Russia would not emphasize nuclear safety and security as much in their nuclear power projects compared to the United States or other suppliers. For example, the United States requires customers of U.S.-origin nuclear reactors to sign Additional Protocols with the IAEA, which are customized safeguards agreements for an individual country's nuclear energy program which impose additional inspection requirements on countries which have them. Russia does not require an IAEA Additional Protocol for its customers.[35] China "is willing to sign opaque cooperation arrangements with emerging nuclear states that allow the transfer of nuclear technology with few nonproliferation restrictions."[36] As autocratic countries, safety and security are prioritized only as long as they serve the purposes of those in authority. Russian troop behavior at Chernobyl and the use of the Zaphorizhzhia power plant as a fortress in Ukraine in 2022 point to a disregard for nuclear safety and security whenever pursuit of another political objective is deemed more important.[37] The Chinese Communist Party's early efforts to cover up outbreaks of Covid-19 is also an example of how public safety is secondary to not embarrassing the government.[38] Neither example instills confidence that nuclear safety and security would be prioritized in the case of an emergency at a Russian or Chinese nuclear power project if that prioritization ran counter to the political interests of their suppliers. The fact that any importing country would be subject to IAEA inspections offers some reassurance of oversight, but ultimately the beliefs and attitudes of those engaged in the day-to-day operation of the plant determine safety and security culture. Countries are going to build

safest way to make electricity, according to study," *Washington Post*, April 2, 2011, https://www.washingtonpost.com/national/nuclear-power-is-safest-way-to-make-electricity-according-to-2007-study/2011/03/22/AFQUbyQC_story.html.

35   Adam N. Stulberg and Jonathan Darsey, "Moving beyond Self-Restraint: Bilateral Commercial Nuclear Supply and US-Russian Tacit Understanding on Nuclear Security and International Safeguards," in *End of an Era: The United States, Russia, and Nuclear Nonproliferation*, eds. Sarah Bidgood and William C. Potter (Monterey, CA: James Martin Center for Nonproliferation Studies, August 2021), 89, https://nonproliferation.org/wp-content/uploads/2021/08/end-of-an-era-the-united-states-russia-and-nuclear-nonproliferation__potter_bidgood_book.pdf#page=102.

36   Ibid.

37   "IAEA Chief Grossi to Head Assistance Mission to Ukraine's Chornobyl Nuclear Power Plant Next Week," International Atomic Energy Agency, April 22, 2022, https://www.iaea.org/newscenter/pressreleases/iaea-chief-grossi-to-head-assistance-mission-to-ukraines-chornobyl-nuclear-power-plant-next-week; Ed Browne, "Russian Troops Throw Zaporizhzhia Staff in Nuclear Plant Basement: Report," *Newsweek*, June 22, 2022, https://www.newsweek.com/russian-forces-ukraine-nuclear-power-plant-staff-basement-zaporizhzhia-1717948; and Hugo Bachega, "Russia using Zaporizhzhia nuclear power plant as army base – Ukraine," BBC, August 8, 2022, https://www.bbc.com/news/world-europe-62469740.

38   Susan V. Lawrence, *COVID-19 and China: A Chronology of Events*, CRS Report No. R46354 (Washington, DC: Congressional Research Service, May 2020), 11, https://crsreports.congress.gov/product/pdf/r/r46354.

nuclear power plants in the future for the many reasons already mentioned in this paper, and it is best if they buy from suppliers with strong nuclear safety and security cultures.

Altogether, the United States has much to lose by staying behind Russia and China in the global nuclear energy industry and much to gain by surpassing them. If it does not enhance its position, the United States could be at a disadvantage in a strategically important industry in the future, countries hostile to its interests could be strengthened, and weaker nuclear safety and security cultures could take root in nuclear energy programs around the world.

CURRENT U.S. ACTIONS AND POLICY RECOMMENDATIONS

Fortunately, the United States has taken action to improve R&D (if not construction) in its domestic nuclear energy industry and enhance its prospects for exporting. The Department of Energy's Advanced Research Projects Agency-Energy (ARPA-E) office funds many initiatives to improve existing power technologies and to speed along bringing ASMRs to market, such as the Converting UNF Radioisotopes Into Energy (CURIE) program, the Modeling-Enhanced Innovations Trailblazing Nuclear Energy Reinvigoration (MEITNER) program, the Generating Electricity Managed by Intelligent Nuclear Assets (GEMINA) program, and the Optimizing Nuclear Waste and Advanced Reactor Disposal Systems (ONWARDS) program.[39] The Advanced Reactor Demonstration Program (ARDP) out of the Department of Energy's Office of Nuclear Energy is also a great start. It awarded $160 million in its first round of awards in October 2020.[40] Congress also directed the Department of Energy to establish a high-assay low-enriched uranium (HALEU) availability program in the Energy Act of 2020, the goal of which is to increase the domestic supply of a fuel likely to be used in many ASMR designs.[41] This financial investment in R&D for ASMRs looks set to continue—the Department of Energy's fiscal year (FY) 2023 budget request included $1.7 billion for the Office of Nuclear Energy, one of the highest requests ever for that office.[42] Finally, the NRC is also working to adapt its design certification review process since ASMR designs are fundamentally different from large LWR designs.[43]

While the U.S. government continues to invest in nuclear energy R&D, it is also attempting to pave the way for increased exports. The Department of State announced its Foundational Infrastructure for the Responsible Use of SMR Technology (FIRST) initiative in April 2021, which is a capacity-

39   "DOE Announces $48 Million For New Program to Securely and Economically Recycle Used Nuclear Fuel," Department of Energy, March 15, 2022, https://arpa-e.energy.gov/news-and-media/press-releases/doe-announces-48-million-new-program-securely-and-economically; "Optimizing Nuclear Waste and Advanced Reactor Disposal Systems," Department of Energy, May 19, 2021, https://arpa-e.energy.gov/technologies/programs/onwards; "Generating Electricity Managed by Intelligent Nuclear Assets," Department of Energy, October 2, 2019, https://arpa-e.energy.gov/technologies/programs/gemina; and "Modeling-Enhanced Innovations Trailblazing Nuclear Energy Reinvigoration," Department of Energy, June 4, 2018, https://arpa-e.energy.gov/technologies/programs/meitner.

40   "U.S. Department of Energy Announces $160 Million in First Awards under Advanced Reactor Demonstration Program," Department of Energy, October 13, 2020, https://www.energy.gov/ne/articles/us-department-energy-announces-160-million-first-awards-under-advanced-reactor.

41   Department of Energy, "Request for Information (RFI) Regarding Planning for Establishment of a Program To Support the Availability of High-Assay Low-Enriched Uranium (HALEU) for Civilian Domestic Research, Development, Demonstration, and Commercial Use," *Federal Register*, 76, 71055 (December 14, 2021), https://www.federalregister.gov/documents/2021/12/14/2021-26984/request-for-information-rfi-regarding-planning-for-establishment-of-a-program-to-support-the.

42   "5 Key Takeaways from the Nuclear Energy FY2023 Budget Request," Department of Energy, June 6, 2022, https://www.energy.gov/ne/articles/5-key-takeaways-nuclear-energy-fy2023-budget-request.

43   "Small Modular Reactors (LWR designs)," Nuclear Regulatory Commission, June 7, 2022, https://www.nrc.gov/reactors/new-reactors/smr.html; and "Guidance and information for developing advanced reactor source term," Nuclear Regulatory Commission, August 24, 2022, https://www.nrc.gov/reactors/new-reactors/advanced/nuclear-power-reactor-source-term.html.

building program for countries around the world looking to start nuclear energy programs, with an emphasis on ASMR deployment.[44] The National Nuclear Security Administration's Advanced Reactor International Safeguards Engagement (ARISE) program is similar to the FIRST program in that it promotes responsible deployment of SMRs abroad, but it specifically has a focus on safeguards. Financially, there have been moves that could enable nuclear exports. For example, in December 2019, Congress reauthorized the Export-Import Bank for seven years and allowed its board to convene without quorum, increasing certainty for projects seeking loans, which is crucial for U.S. nuclear companies.[45] Then, in July 2020, the U.S. International Development Finance Corporation ended its ban on financing nuclear projects.[46] And in the case of Romania, which may be the first customer in the world of NuScale's SMR power plant, the U.S. Trade and Development Agency (USTDA) funded its site evaluation process, which was finished in May 2022 with the announcement of the city of Doicesti as the preferred site.[47] In June 2022, President Biden announced the Partnership for Global Infrastructure and Investment effort, which includes the flagship Front-End Engineering and Design (FEED) study to maintain the momentum of constructing a NuScale power plant in Romania. It is jointly funded by the Romanian government's Nuclearelectrica and NuScale.[48]

POLICY RECOMMENDATIONS

The U.S. government's movement on this front in encouraging, but the numbers in Table 1 still show that the United States is behind Russia and China despite these efforts. The three areas that could push the United States forward the most are financing, continuing capacity building abroad, and leading by example by rejuvenating the U.S. domestic nuclear energy sector.

First, financing could be the crucial factor that leads countries who might otherwise choose U.S. suppliers to choose Russian or Chinese ones instead. Because Russia and China both have state-owned nuclear energy corporations, they are able to offer comprehensive packages to potential customers that include financing, often through state-backed loans but other financing models as well, such as build-own-operate. Although it is likely these favor China and Russia more than the potential customers, those customers may not have another choice if other suppliers do not offer financing at all. The U.S. government should make it easier for countries to finance nuclear power projects from U.S. suppliers. That means making it easier to qualify for loans from the U.S. Export-Import Bank for nuclear power projects but also enabling a continuance and expansion of projects such as the USTDA-funded site evaluation for a NuScale plant in Romania.

<hr>

44 "Media Note: Program to Create Pathways to Safe and Secure Nuclear Energy Included in Biden-Harris Administration's Bold Plans to Address the Climate Crisis," Department of State, April 27, 2021, https://www.state.gov/program-to-create-pathways-to-safe-and-secure-nuclear-energy-included-in-biden-harris-administrations-bold-plans-to-address-the-climate-crisis/.

45 U.S. Congress, House, *Further Consolidated Appropriations Act*, H.R.1865, 116th Cong., 1st sess., introduced March 25, 2019, https://www.congress.gov/bill/116th-congress/house-bill/1865/text.

46 "DFC Modernizes Nuclear Energy Policy," U.S. International Development Finance Corporation, July 23, 2022, https://www.dfc.gov/media/press-releases/dfc-modernizes-nuclear-energy-policy.

47 "Nuclearelectrica and NuScale sign MoU, site chosen for SMR," World Nuclear News, May 24, 2022, https://www.world-nuclear-news.org/Articles/Nuclearelectrica-and-NuScale-sign-MoU,-site-chosen; and "IAEA Team in Romania Concludes First Ever Site and External Events Design (SEED) Review for a Small Modular Reactor," IAEA, September 15, 2022, https://www.iaea.org/newscenter/pressreleases/iaea-team-in-romania-concludes-first-ever-site-and-external-events-design-seed-review-for-a-small-modular-reactor.

48 "Media Note: United States Takes Next Step in Supporting Innovative Clean Nuclear Technology in Europe," Department of State, June 26, 2022, https://www.state.gov/united-states-takes-next-step-in-supporting-innovative-clean-nuclear-technology-in-europe/.

Second, the U.S. government should continue and even increase its current capacity-building efforts in countries looking to start nuclear energy programs. Starting a national nuclear energy program takes decades of high-level political engagement, and nuclear newcomers appreciate help in this monumental task. Nuclear newcomers must write and pass legislation and regulations, create new government agencies (e.g., a nuclear regulatory agency), evaluate reactors and sites, and sign intergovernmental agreements. The IAEA's Milestones Approach program is helpful for nuclear newcomer countries, but U.S. efforts such as the FIRST program both supplement the IAEA's program and increase the likelihood countries will select U.S. suppliers by deepening strategic ties. FIRST is a great step in the right direction and should be expanded upon. If possible, Section 123 Agreements with countries likely to start nuclear energy programs should even be drafted in advance so that they are ready to be introduced in Congress when the time comes for these nuclear newcomer countries to select suppliers. The goal is to make choosing a U.S. supplier the easiest, clearest choice.

Finally, the United States should lead by example domestically and rejuvenate its own nuclear energy sector, which has atrophied since the 1970s. This is easier said than done. While the federal government has taken steps to invest in the domestic nuclear energy industry in terms of R&D and paving the way for exports, actually building new nuclear power plants domestically would be challenging on multiple fronts. The United States has practically no new nuclear construction, so the skills and supply chains needed to build nuclear power plants have largely been lost. For example, existing plants are often shut down years before their operating licenses expire. (14 reactors have been shut down prematurely since 2013).[49] The sector's difficulty is partly because the U.S. public remains largely skeptical of nuclear power despite its safe operation record, but nuclear energy is stunted in the United States mainly for economic reasons. Low electricity prices disincentivize capital-intensive projects such as nuclear power plants, and the long construction times make it difficult to predict what electricity prices will be when the plant becomes operational. As the International Energy Agency reported in its 2019 report *Nuclear Power in a Clean Energy System*, "markets and regulatory systems often penalize nuclear power by not pricing in its value as a clean energy source and its contribution to electricity security."[50] Fossil fuels, nuclear energy's main competitor, are also artificially cheap because their harm to the environment is not priced in. A carbon tax and tax incentives for zero-carbon-emission energy production are most likely necessary but not sufficient conditions for a domestic nuclear energy renaissance. A nuclear energy version of the Creating Helpful Incentives to Produce Semiconductors for America (CHIPS) Act could be a first step for the domestic nuclear energy industry.[51] This could allow the United States to lead by example in decarbonizing its electricity grid, and a strong domestic market would strengthen its nuclear energy suppliers in the export market as well.

49   International Energy Agency, *Nuclear Power in a Clean Energy System* (Paris: May 2019), https://www.iea.org/reports/nuclear-power-in-a-clean-energy-system; and "Initial License Renewal Filings For U.S. Nuclear Power Plants," Nuclear Energy Institute, May 2021, https://www.nei.org/resources/statistics/us-nuclear-license-renewal-filings; "Knowledge Loss Risk Management in Nuclear Organizations," IAEA, Nuclear Energy Series No. NG-T-6.11, 3–4, https://www-pub.iaea.org/MTCD/Publications/PDF/17-35661_PUB1734_web.pdf; Mark Holt and Phillip Brown, *U.S. Nuclear Plant Shutdowns, State Interventions, and Policy Concerns*, CRS Report no. R46920 (Washington, DC: Congressional Research Service, June 2021), 4, https://crsreports.congress.gov/product/pdf/R/R46820/3; and Mitchell Boatman, "Palisades owner says 'number of hurdles' need to be cleared to reopen plant," *The Holland Sentinel*, September 12, 2022, https://www.hollandsentinel.com/story/news/environment/2022/09/12/palisades-owner-number-of-hurdles-to-clear-before-possible-reopening/69486660007/.

50   International Energy Agency, *Nuclear Power in a Clean Energy System* (Paris: IEA, May 2019), 4, https://www.iea.org/reports/nuclear-power-in-a-clean-energy-system.

51   U.S. Congress, House, *CHIPS for America Act*, 116th Cong., 2nd sess., introduced June 11, 2020, https://www.congress.gov/bill/116th-congress/house-bill/7178.

CONCLUSION

Nuclear technology is powerful whether it is used for destructive or peaceful applications, so it is simultaneously positive that a global nuclear energy renaissance is gaining momentum but imperative that the global nuclear energy sector grows responsibly. The United States has fallen behind China and Russia in projecting influence via exports. This is a strategic liability for the United States in that (1) it may be at a disadvantage in an important future industry, (2) China and Russia will only grow stronger by dominating that same industry, and (3) China and Russia will influence nuclear safety and security cultures in the countries they supply. While the United States has made an effort to better its position by investing in R&D and paving the way for nuclear exports, the United States is still behind China and Russia. To improve its position, the U.S. government should bolster its financing options for potential customers of U.S. reactors, continue and amplify its capacity-building efforts in nuclear newcomer countries, and lead by example in rejuvenating its own domestic nuclear energy sector. By doing so, it will be able to maintain its position and influence in humanity's ongoing effort to harness the peaceful power of the atom while avoiding its destructive potential.

Russian Thought, Global Challenges

A Dive into Russian Perspectives of Strategic Stability

By Brandon Cortino[1]

INTRODUCTION

Russia and the United States are on a geopolitical collision course owing to mutual distrust and incompatible visions of the international order. The current Ukraine crisis is a symptom of deep Russian insecurities and demonstrates Moscow's efforts to wrest geopolitical control of its "sphere of influence" in a manner consistent with the Yalta agreement, and similarly demonstrates the Kremlin's insecurity about its current borders with a Western military alliance.[2] Russian foreign minister Sergei Lavrov has previously stated that the encroachment of the North Atlantic Treaty Organization's (NATO) military infrastructure is a threat to Russian national security.[3] There has been considerable research on the U.S. perception of Russian goals that suggests Russian international relations use a model like realism, but little academic literature and policymaking adequately reviews Russian perceptions.

This paper will review strategic stability from the Russian lens using Russian-language sources, examine conflicts from the U.S. perspective, and search for a cooperative middle ground to raise the strategic threshold. The paper will first review Russia's shift in disposition and then compare Russian and English linguistic terminology, including differences in each language's understanding of deterrence. The paper will also compare Russian and U.S. strategic cultures and make recommendations based on the historic conditions leading to the current political environment.

---

1   Brandon Cortino is a nuclear security specialist who prioritizes bridging the communications gap between Russia and the United States through a cultural and linguistic perspective after living abroad in Russia.

2   Angela Stent, Putin's World: Russia Against the West and with the Rest (New York: Twelve Publications, 2019).

3   "Russia Says U.S. Actions Threaten its National Security," Reuters, October 9, 2016, https://www.reuters.com/article/us-russia-usa-idUSKCN1290DP.

In answering these questions, an underlying assumption exists within the United States that the U.S. foreign policy community understands Russian security concerns. This paper argues that even if Russia is understood, appropriate policy actions are not undertaken. Through Western eyes, the Ukraine crisis has been lambasted as completely irrational and avoidable, with scholars decrying Vladimir Putin as a bad strategist.[4] In Putin's eyes, the Ukraine crisis has been a trainwreck 14 years in the making, as Russia perceives NATO expansion as a threat reminiscent of Napoleon and Hitler burning Moscow.[5] At the 2008 Bucharest Memorandum, where President George W. Bush announced plans to eventually welcome Ukraine and Georgia into NATO, Putin dismissed the Ukrainians' right to self-determination to President Bush, saying that Ukraine was not a sovereign country, and later warned of stopping NATO expansion one way or another.[6] Putin later indicated that NATO expansion was a direct threat to Russian national security, reflected in the 2014 military doctrine as the first item on the list of foreign threats as well as in Russia's 2016 foreign policy concept.[7]

The consequences of Russia's geopolitical insecurity surrounding NATO expansion include three separate military conflicts of escalating intensity: with Georgia in 2008, Crimea in 2014, and the rest of Ukraine in 2022. The United States and NATO state that the alliance is a defensive organization, but as John Mearsheimer points out: "The issue at hand is not what Western leaders say NATO's intentions are; it is how Moscow sees NATO's actions."[8] The Soviet Union and Russia have persistently viewed NATO as a threat, but different leaders have employed different security strategies to address this insecurity.[9]

Mikhail Gorbachev was famous for ameliorating tensions between East and West through arms control and was a champion of cooperation, working to make international relations match domestic reforms. Boris Yeltsin was willing to cooperate with the West on a limited front but was hesitant to agree to NATO expansion into the Russian sphere of influence, going so far as to send President Bill Clinton a letter expressing Russian opposition to the Czech Republic and Poland joining NATO.[10] During Yeltsin's tenure, the perspective of Russian society shifted into believing that the United States was a threat, which was reflected in Yeltsin's foreign policy shifting from warm to hesitant, fueled by Yevgeny Primakov's insistence on the country's return to great power status.[11] In 1997, Russian defense minister

---

4   Joshua Rovner, "Putin's Folly: A Case Study of an Inept Strategist," War on the Rocks, March 16, 2022, https://waronth-erocks.com/2022/03/putins-folly-a-case-study-of-an-inept-strategist/; and Edward Geist, "Is Putin Irrational? What Nucle-ar Strategic Theory Says About Deterrence of Potentially Irrational Opponents," RAND Corporation, March 8, 2022, https://www.rand.org/blog/2022/03/is-putin-irrational-what-nuclear-strategic-theory-says.html.

5   An argument could be made that this was a problem 31 years in the making since the collapse of the Soviet Union; The Russian Ministry of International Affairs posted a picture for the May holidays with four helmets in a row: a Crusader helmet, a Lithuanian helmet, a French officer's hat, and a World War II German infantry helmet. There is a fifth blank spot implying NATO is next. The picture's caption translates to: "We politely asked them not to expand eastwards."

6   David Remnick, "Putin's Pique," New Yorker, March 10, 2014, https://www.newyorker.com/magazine/2014/03/17/putins-pique.

7   "The Russian Federation's Military Doctrine," Russian Federation's Security Council, December 25, 2014, http://www.scrf.gov.ru/security/military/document129/; Vladimir Putin, "The Russian Federation's Foreign Policy Concept," Russian Federation's Security Council, November 30, 2016, http://www.scrf.gov.ru/security/international/document25/; and Steven Erlanger, "Putin, at NATO Meeting, Curbs Combative Rhetoric," New York Times, April 5, 2008, https://www.nytimes.com/2008/04/05/world/europe/05nato.html.

8   John Mearsheimer, "John Mearsheimer on Why the West is Principally Responsible for the Ukrainian Crisis," The Economist, March 19, 2022, https://www.economist.com/by-invitation/2022/03/11/john-mearsheimer-on-why-the-west-is-principal-ly-responsible-for-the-ukrainian-crisis.

9   Andrei P. Tsygankov, Russia's Foreign Policy: Change and Continuity in National Identity (Lanham, MD: Rowman & Littlefield, 2019).

10   Roger Cohen, "Yeltsin Opposes Expansion of NATO in Eastern Europe," New York Times, October 2, 1993, https://www.nytimes.com/1993/10/02/world/yeltsin-opposes-expansion-of-nato-in-eastern-europe.html.

11   Tsygankov, Russia's Foreign Policy.

Primakov met with his Ukrainian counterpart to discuss borders and mentioned Yeltsin's unhappiness with Ukraine for conducting naval exercises with NATO.[12] Dmitry Medvedev pursued a "reset" with the United States, pledged helicopters to NATO for assistance in Afghanistan, renewed bilateral military cooperation under the Defense Working Group of the Bilateral Presidential Commission, and agreed to an arms control treaty.[13] In contrast, Putin has frozen the relationship.

STRATEGIC STABILITY AND RUSSIA'S TURN FROM THE WEST

This paper argues that differing perceptions regarding NATO's role in Europe are in part forged by cultural and linguistic peculiarities. What is encompassed in strategic stability depends on a country's perception and history.[14] Alexei Arbatov writes that when the START I treaty was signed, both the United States and the Soviet Union had differing definitions of strategic stability, with the Soviet Union viewing it as a "prism of ratio of strategic offensive and defensive arms."[15] Years after the collapse of the Soviet Union, the concept behind this article and its arguments is still vague.

In an exclusive interview for this paper, the United States' contemporary definition of strategic stability was explained by U.S. Strategic Command's deputy commander. Strategic stability is divided into three layers: the first layer removes first-strike incentives, the second layer seeks to prevent direct war between nuclear powers, and the third ideal layer is for strategic stability to act as a framework to avoid conflicts between nations.[16] When START I was signed, arms control was effectively ratified as a central tenet of strategic stability.

However, the Russian definition of strategic stability is considerably broader. The Russian understanding of this term immediately expanded after START I to include any variable impacting national stability on an existential level.[17] After the collapse of the Soviet Union, Russia was in the uncomfortable position of needing to redefine its fundamental existence and destiny.[18] It is only logical that a country experiencing concurrent crises would reevaluate its international strategy from a survival lens.[19] This effect was exacerbated by the "policy drift" between Gorbachev's cooperative Western-oriented administration and the traditionalist Putin regime.

12   A. Smirnov, "So Far the Border Between Ukraine and Russia is Being Drawn Only On Paper: A Statement is Signed on Treaty-Based Legal Delineation of Borders," *Current Digest of the Post-Soviet Press* 49, no. 35 (October 1997).

13   "U.S.-Russia Relations: 'Reset' Fact Sheet," The White House (Obama), June 24, 2010, https://obamawhitehouse.archives.gov/the-press-office/us-russia-relations-reset-fact-sheet.

14   In 2017, Colonel Mike Shinners published an article for the NATO Joint Warfare Center that states the following: "However, differences in national cultures and communication norms may affect us when trying to reach other cultures within NATO." This premise is true for the largely democratic nations within NATO and is doubly true for a country as culturally distinct as Russia outside of NATO. "Communication, Culture and Effective Teams," Joint Warfare Center, NATO, April 17, 2017, https://www.jwc.nato.int/articles/communication-culture-and-effective-teams.

15   Alexei G. Arbatov et al., *Strategic Stability after the Cold War* (Moscow: Institute of World Economy and International Relation, Russian Academy of Science, 2010), https://www.files.ethz.ch/isn/144185/10011.pdf.

16   Thomas Bussiere, interview by author, June 27, 2022.

17   Arbatov et al., *Strategic Stability after the Cold War.*

18   Jeffrey Mankoff, *Russian Foreign Policy: The Return of Great Power Politics* (Lanham, MD: Rowman & Littlefield Publishers, 2009).

19   Rebecca Friedman Lissner, "What is Grand Strategy? Sweeping a Conceptual Minefield," *Texas National Security Review* 2, no. 1 (2018), doi:10.26153/tsw/868; Lawrence Rubin and Adam N. Stulberg, eds., *The End of Strategic Stability?: Nuclear Weapons and the Challenge of Regional Rivalries* (Washington, DC: Georgetown University Press, 2018), https://www.jstor.org/stable/j.ctv75db5d; and Nina Silove, "Beyond the Buzzword: The Three Meanings of 'Grand Strategy,'" *Security Studies* 27, no. 1 (2018): 27–57, doi:10.1080/09636412.2017.1360073.

Gorbachev was convinced of the necessity to cooperate with the West as a continuation of reforms within Soviet society.[20] Putin views these same Western countries as interlopers and has frequently warned against continued NATO expansion, as demonstrated by repeating the talking point that NATO promised not to expand past Germany. Robert Zoellick, the lead U.S. negotiator for the 2+4 reunification dialogue, stated in a panel on January 1, 2022, that Secretary of State James Baker asked Gorbachev if he would prefer an independent unified Germany or a reunified Germany associated with NATO not one inch farther east.[21] After a German referendum overwhelmingly supported reunification, Gorbachev agreed to allow Germany to join NATO on the condition that no new NATO infrastructure would be placed in East Germany.[22] Gorbachev confirmed that the topic of NATO expansion did not come up at all, suggesting that it was not on the minds of Soviet leadership while discussing Germany's future, nor did it seem to have been the case for the West.[23] The "expansion promise" was only made in the context of NATO infrastructure. However, NATO infrastructure and NATO membership are different things. Secretary of Defense Robert Gates stated in 2000 that Gorbachev and others were misled about NATO's intentions to expand, which further exacerbated Russia's frustration following the Soviet Union's collapse.[24] Even Gorbachev's cooperative demeanor shifted, and he outright accused NATO of trying to build a new empire in a December 2021 op-ed in *Kommersant*.[25]

There was no switch flipped once Putin became president but rather a gradual shift underpinned by anathema toward NATO, with the perception that the West was exploiting Russia's weakness following the Soviet Union's disintegration.[26] This hesitation was strengthened by NATO's 1999 bombing campaign in Belgrade. In response to NATO's bombing, Prime Minister Primakov reversed course back to Russia while flying to the United States. At the same time, the Russian media presented coverage of Belgrade that united Russians against what they perceived as a common threat facing a fellow Slavic nation.[27] The U.S. bombing was viewed as barbaric and terrifying by Russians of all political stripes, with a poll conducted in 1999 finding 63 percent of Russian citizens believing NATO and the United States could be a threat to Russia.[28] Some of this concern stemmed

20   Mikhail Gorbachev, "Commanding Heights: Interview with Mikhail Gorbachev," PBS, April 23, 2001, https://www.pbs.org/wgbh/commandingheights/shared/minitextlo/int_mikhailgorbachev.html.

21   Robert Zoellick, "Setting the Record Straight on NATO Enlargement," (public event, Wilson Center, Washington, DC, January 31, 2022), https://www.wilsoncenter.org/event/setting-record-straight-nato-enlargement-interview-robert-zoellick-us-lead-negotiator-24.

22   Ibid.; and Maxim Korshunov, "Mikhail Gorbachev: I am Against All Walls," Russia Beyond the Headlines, October 16, 2014, https://www.rbth.com/international/2014/10/16/mikhail_gorbachev_i_am_against_all_walls_40673.html.

23   Pavel Palazhchenko, "Mikhail Gorbachev and the NATO Enlargement Debate: Then and Now," in *Exiting the Cold War*, edited by Daniel S. Hamilton and Kristina Spohr (Washington, DC: Brookings Institution, 2019), 443.

24   Robert Gates, "Oral History," Miller Center, July 24, 2000, https://millercenter.org/the-presidency/presidential-oral-histories/robert-m-gates-deputy-director-central. In Vladimir Putin's 2005 State of the Nation address, he said: "Above all it should be acknowledged that the collapse of the Soviet Union was the largest/greatest geopolitical catastrophe of the century. For the Russian people it because a real drama. Tens of millions of our citizens ended up outside Russian territory's borders. This epidemiological collapse also affected Russia itself." The use of the terms "catastrophe" and "epidemic" from a relatively restrained public speaker such as Putin signify deep emotions that have arguably not been properly weighed in policy considerations. "Message to the Federal Assembly of the Russian Federation," President of Russia, April 5, 2005, http://kremlin.ru/events/president/transcripts/22931.

25   Mikhail Gorbachev, "Gorbachev on NATO's expansion plans: they've been struck in the head with arrogance." *Kommersant*, December 24, 2021, https://www.kommersant.ru/doc/5143543.

26   Maria Lipman, "Putin's World: Russia Against the West and With the Rest," *Foreign Affairs*, September/October 2019, https://www.foreignaffairs.com/reviews/capsule-review/2019-08-12/putins-world-russia-against-west-and-rest.

27   Ivan Krastsev, "Экспериментальная родина. Разговор с Глебом Павловским" [Experimental Homeland. A Conversation with Gleb Pavlovski], *Evropa*, 2018.

28   "Survey: Two-Thirds of Russians fear NATO Attack," CNN, April 3, 1999, http://edition.cnn.com/WORLD/europe/9904/03/kosovo.russia/.

from the way Russian media presented the bombings, but it was ultimately a zeitgeist moment in the Russian drive to protect itself.

Alongside the beginning of the second Chechen war and a rash of terrorism in Russia, Belgrade's bombing led Yeltsin to find "not a young reformer, not a bespectacled technocrat, but a strong man in uniform."[29] Yeltsin's decision to find a strong candidate from within the security apparatus, such as the fictional spy Stierlitz, began the cascade of events leading to Russia homesteading the political wilderness in the twenty-first century.[30] Russia's turn against the West was fueled by domestic skepticism and Putin's cynicism about the world order in his early years as president. Perhaps the earliest clear example was President Bush's unilateral withdrawal after 9/11 from the Anti-Ballistic Missile (ABM) Treaty to develop potential countermeasures against future terror attacks. Putin saw Bush's sudden abandonment of the treaty as a risk to the Russian nuclear deterrent's effectiveness, as evidenced by later commissioning the hypersonic weapon platform now known as the SSC-8, but he also viewed the withdrawal as an insult. In a press conference at the Bundestag on September 25, 2001, Putin said:

> In spite of all of the positive achievements of the past decades, we have not yet developed an efficient mechanism for working together. The coordinating agencies set up so far do not offer Russia real opportunities for taking part in drafting and making decisions. Today decisions are often made, in principle, without our participation, and we are only urged afterwards to support such decisions. After that they talk again about loyalty to NATO. They even say that such decisions cannot be implemented without Russia. Let us ask ourselves: is this normal? Is this true partnership?[31]

Context for Putin's frustration can be found in an article by the Russian former minister of foreign affairs Igor Ivanov. After 9/11, Ivanov had a meeting with Secretary of State Condoleezza Rice where she informed him that the United States unilaterally decided to withdraw from the ABM Treaty and requested Russia to sign a document, without consulting Russia as an equal party.[32] No country likes being treated as a junior partner, and as Putin's speech at the Bundestag demonstrates, Putin was unhappy with the United States making major arms control decisions without involving Russia beyond a symbolic level.

DETERRENCE OR *SDERZHIVANIE*?

The term "deterrence," stemming from the Latin "to terrify or discourage away from," is separate from the approximately equivalent Russian term "*sderzhivanie*," translating as "to hold back/contain." The closest Russian term to deterrence is "*ustrashenie*," which translates as "nuclear intimidation," but it is used to imply illegitimate nuclear blackmail by the United States. Russian forces overwhelmingly use *sderzhivanie* instead of associating with the negative implications that *ustrashenie* conjures.[33] The Russian term *sderzhivanie* is broader than the American parallel, encompassing coercive statecraft,

29   Krastsev, "Experimental Homeland."
30   Max Otto Von Stierlitz is a fictional spy roughly equivalent to James Bond. The character was designed to rehabilitate the image of intelligence officers into an attractive and patriotic image in the post-Stalin era for recruiting and propaganda purposes.
31   Vladimir Putin, "Speech in the Bundestag of the Federal Republic of Germany," (speech, Bundestag, Berlin Germany, September 25, 2001), http://en.kremlin.ru/events/president/transcripts/21340.
32   "Putin Went into a Death Spiral: The Secret Preamble to the Blowout with the West," *Moskovskiy Komsomolets*, March 15, 2022, https://www.mk.ru/politics/2022/03/15/putin-voshel-v-mertvuyu-petlyu-sekretnaya-predystoriya-razryva-s-zapadom.html.
33   Kristin Ven Bruusgaard, "Russian strategic deterrence," *Survival* 58, no. 4 (2016), doi:10.1080/00396338.2016.1207945.

compellence, and intra-war deterrence.[34] While the American term is centered around generating fear in an adversary's mind, the Russian term's elements of "contain/restrain" does not focus on such connotations and is expedient about limiting an adversary's existing movement.

The distinction in how these respective terms are used is perhaps the most crucial component. Deterrence is used to prevent an adversary from engaging in a hypothetical action and targets policymakers while existing as a latent consequential threat. *Sderzhivanie* is used in peace and wartime to restrain adversaries from additional action by targeting society at large as well as leaders. The targeting difference reflects respective U.S. and Russian societies and theories of international relations. Understanding and adjusting for differences that are not immediately obvious leads to more nuanced policy that improves overall strategic stability.

The definition of *sderzhivanie* has grown in the post-Cold War era. When the term originally appeared in the Russian security lexicon, it referred exclusively to nuclear deterrence. In 2014, *sderzhivanie* grew conceptually to include "pre-nuclear *sderzhivanie*," defined as "the suite of foreign policy, military, and military-technical measures directed at the prevention of aggression against the Russian Federation by non-nuclear means."[35] Strategic *sderzhivanie* refers now to the principles of deterrence, containment, coercion, and compellence. Perhaps the most crucial implication of strategic *sderzhivanie*, according to Samuel Charap's research, is how destabilizing such a broad concept can be. If Russia views itself in a persistent state of defensive and reactionary intra-war conflict with a massive array of capabilities to utilize, there is considerable risk of Russia using capabilities in an overly escalatory manner without realizing what their counterpart's perception will be. This risk is magnified because Russian *sderzhivanie* does not have as strong of an institutional emphasis on adversary psychology as American deterrence.[36] Although the United States must work on its cultural approach to handling Russia, the same can be said of Russia's approach to the United States.

CULTURAL DIMENSIONS

There are a series of useful cultural metrics designed by Geert Hofstede to help tabulate differences between cultures.[37] This system provides limited insight into Russian and American perspectives but is a useful tool for quickly identifying and explaining driving cultural factors. This paper uses Hofstede's cultural dimensions because it is a relatively well-known system and has a high degree of generalizability. Russia has extremely high uncertainty avoidance.[38] For the purposes of strategic planning, having high uncertainty avoidance means that certainty is everything, and every possible beneficial action will be pursued despite higher known costs.

Uncertainty avoidance can be extrapolated to ongoing circumstances with Russia's concerns over NATO expansion. Prior existential attacks on Russian statehood by Napoleon and Hitler shaped tolerance for foreign military presence and infrastructure. In the minds of many Russian *siloviki*, the

---

34   Samuel Charap, "Strategic Sderzhivanie: understanding contemporary Russian approaches to 'deterrence,'" George C. Marshall European Center for Security Studies, September 2020, https://www.marshallcenter.org/en/publications/security-insights/strategic-sderzhivanie-understanding-contemporary-russian-approaches-deterrence-0.

35   "Военная доктрина Российской Федерации, [The Military Doctrine of the Russian Federation]," Russian Federation, December 30, 2014, https://rg.ru/2014/12/30/doktrina-dok.html.

36   Charap, "Strategic Sderzhivanie."

37   Geert Hofstede, "Dimensionalizing Cultures: The Hofstede Model in Context," *Online Readings in Psychology and Culture* 2, no. 1 (2011): 2307–0919, doi:10.9707/2307-0919.1014.

38   Denise Rotondo Fernandez et al., "Hofstede's country classification 25 years later," *Journal of Social Psychology* 137, no. 1 (1997): 43–54, doi:10.1080/00224549709595412.

Russian defense industry's influential figures, the biggest difference between NATO and previous adversaries is that NATO just has not attacked yet. Russian training exercises and military doctrine center around four regions represented by the cardinal directions of west, north, east, and south. These training exercises include forces from each region to support an "all of Russia" approach meant to complete multiple simultaneous military objectives on a large scale, demonstrating the willingness to invest in higher "known" costs to mitigate unknown worst-case scenarios.[39]

Although the Russian military is divided in four regions, this paper's focus is on the Russia-NATO relationship. Beyond a shared history, Russia perpetually feels most vulnerable on its western front, which has no natural defensive lines, a considerably smaller "buffer zone" compared to 35 years ago, and scars from Operation Barbarossa still within living memory for the population. Western analysts often dismiss the depth of Russian pain inflicted by the Nazis annihilating many cities west of the Urals. Approximately 20 to 27 million Soviet citizens died, the majority of them Russian. Nearly every Russian family lost members to the war, and the memory remains fresh.[40]

PITFALLS AND RECOMMENDATIONS

In an increasingly frosty relationship with ugly conflicts, it is crucial to find common ground and avoid pitfalls. Failure to collaborate introduces consequences affecting the entire global community, as evidenced by the second- and third-order effects of the invasion of Ukraine. This includes internal impacts such as the new wave of Russian emigrants hoping to avoid the economic fallout of conflict and the risks of conscription who are taking their money and expertise to other countries. On a regional level, the European Union is taking concrete steps to limit energy imports from Russia, which will impact Russia's economic health and influence in Europe while also risking the wellbeing of EU citizens during the cold winter months. Worse still is the threatened humanitarian toll of skyrocketing grain prices, which places poorer global regions at risk.

The United States and Russia share the core objective of existential survival, allowing for limited cooperation on urgent matters, but strategic priorities differ. The United States is comfortable with its role in the world as the global hegemon, but Russians are generally highly risk averse and perceive NATO's expansion as a direct threat owing to prior invasions from Europe that define the Russian worldview. Although Zbigniew Brzezinski presented the Russian population as relatively indifferent to NATO expansion, almost every facet of the Russian political world strongly opposed it.[41] At the time of publication, a greater diversity of views and lifestyles was respected in Russia, but they were virtually all united by intense opposition to NATO. In a very real sense, the Soviet Union was the most recent empire to collapse. Sensibilities of what land "belongs" to Russia extends far beyond its borders into the Baltic states, the Caucasus, and Ukraine, as stated by Putin, as believed by a considerable portion of the Russian population, and as reflected in Russia's grand strategy of "three levels of war," referring to the post-Soviet near abroad, regional theater conflicts, and total war. So long as Russia follows its existing political trajectory and the West expands and welcomes new partners that encroach on Russian views of territory in their sphere of influence, there will continue to be struggles across the spectrum of conflict in the name of geopolitics and maintaining a defensive bubble.

---

39   Stephen R. Covington, *The Culture of Strategic Thought Behind Russia's Modern Approaches to Warfare* (Cambridge, MA: Harvard Kennedy School, Belfer Center for Science and International Affairs, 2016), https://www.belfercenter.org/sites/default/files/legacy/files/Culture%20of%20Strategic%20Thought%203.pdf.

40   The same can be said of Ukrainian families.

41   Joseph L. Black, "Russia and NATO expansion eastward: Red-lining the Baltic states," *International Journal* 54, no. 2 (1999): 249–266, doi:10.2307/40203375.

To avoid agitating Russia, the United States at a minimum should temporarily stop discussing Ukraine joining NATO. Gorbachev went on record stating opposition, going so far as to call NATO a new empire, Yeltsin penned a letter to President Clinton opposing NATO expansion, and Putin has made his intention to stop Ukraine from joining NATO painfully clear. When such radically different leaders agree on one foreign policy position, it bears acknowledgement. As Anna Louise Strong wrote: "The Russians never wanted to be a mystery. When I first went to Moscow … they were explaining themselves in tomes of Marxist logic to all who would hear. Most people wouldn't; they called the explanations propaganda. … After a while the Russians stopped explaining. They let their actions speak."[42] This is a line from a book written in 1941, and it unfortunately still reflects foreign policy challenges. Although U.S. diplomats state that no third-party country determines who joins NATO, few members of the alliance would willingly admit a country in ongoing armed conflict, which Putin understands. Former U.S. ambassador to Russia Michael McFaul has suggested that the United States acted in bad faith regarding Ukraine's NATO candidacy to frustrate Russian ambitions.[43] Proposing a NATO membership moratorium is hypothetically possible, but Russian foreign minister Sergei Lavrov has stated such an idea is not appealing.[44] NATO expansion is almost universally perceived as a threat in Russia, and in the interest of improving strategic stability, an unofficial pause may be necessary to mitigate further spiraling outside of Ukraine.

In the interest of maintaining strategic stability, the three major nuclear powers should commit to more nuanced top-down decisionmaking. Anna Peczeli and Ben Bahney suggest employing observation officers from China, Russia, and the United States at major nuclear command, control, and communications (NC3) facilities.[45] The United States already strives to avoid mirror imaging with a broad array of specialists reviewing strategic stability issues considering U.S., allied, and adversary perspectives.[46] This should be further prioritized and include senior leadership from affected countries. However, this is a responsibility for Russia and China as well, and given Russia's current political path, improvements to strategic stability do not appear imminent.

Further, it is the responsibility of both Russia and the United States to avoid inducing conflicts in other countries in the interest of geopolitical competition and engage in good-faith efforts to understand the other side's perspective. One factor hindering attaining a higher level of understanding is that the Russian and U.S. governments have both overly complicated the hiring process for positions within their respective diplomatic sectors due to concerns surrounding insider threats. This failure to reward curiosity and knowledge about other major powers results in a feedback loop of mirror imaging and myopic hawkishness akin to the Cold War. It is extremely difficult to develop a subject matter expert who has never visited their target country, and diplomatic efforts suffer for it. Policymakers are ethically obligated to listen to foreign policy experts to promote healthy and beneficial relationships in the interest of peace and stability.

---

42   Anna Louise Strong, *The Soviets Expected It* (New York: Dial Press; Toronto: Canadian Tribune Publishing, 1941).
43   Michael McFaul, Twitter post, May 22, 2022, 8:56 a.m., https://twitter.com/McFaul/status/1528404483016822784.
44   "As War Looms Larger, What are Russia's Military Options in Ukraine?" *The Economist*, January 22, 2022, https://www.econo-mist.com/europe/what-are-russias-military-options-in-ukraine/21807240.
45   A. Peczeli and B. Bahney, *The Role of Nuclear-Conventional Intermingling on State Decision-making and the Risk of Inadvertent Escalation*, Lawrence Livermore National Laboratory, October 5, 2021, LLNL-TR-827542, https://www.osti.gov/servlets/purl/1836194.
46   Bussiere, interview by author, June 27, 2022.

CONCLUSION

Russia and the United States have differing conceptualizations of terms relevant to international relations that risk leading to misunderstandings because of differing underlying assumptions. To better negotiate and maintain peace, these gaps should be continually evaluated. The Russian principle of *sderzhivanie* is broader and more inclusive than the approximate American equivalent of deterrence. *Sderzhivanie* is based around the idea of containing or restraining adversaries at any cost across the spectrum of conflict, while deterrence is a largely psychological concept.

Cultural dimensions provide a window into how uncertainty avoidance is reflected in policy. Russia has extremely high uncertainty avoidance owing to collective trauma from Napoleon's invasion and Operation Barbarossa during World War II. Western analysts habitually underestimate the depth of the Great Patriotic War's impact on the Russian psyche, and this is representative of a deeper problem in understanding the Russian perspective. Russia holds a generational opposition to NATO expansion, as demonstrated by every leader since Gorbachev speaking strongly against the idea. To preserve the strategic relationship, the United States should stop discussing further NATO expansion at this time, introduce more experts with on-the-ground experience in international relations, and look for more ways to facilitate communication.

Safety in the Next Generation of Nuclear Reactors

Fusion and the Risks of Nuclear Proliferation

By Daniel Gum[1]

ABSTRACT

Long considered a technology that was perpetually 20 years away, recent advances in the science and engineering of fusion power plants have enabled a renaissance of sorts, with the number of private companies pursuing fusion doubling within the last decade. Their approach to solving the clean energy problem is characterized by a wide variety of methods and techniques, all of which currently lack the strong oversight imposed by the International Atomic Energy Agency (IAEA) and domestic governing bodies. Many of these private companies are aiming for pilot plants within the next decade, with intent to begin major commercial operation in the 2040s and 2050s. With this rapid growth in mind, this paper reviews the development of commercial fusion; how these power plants could be used to make nuclear weapons in covert, clandestine, and breakout scenarios; and the proliferation risks from specific approaches to fusion. The paper further discusses which state or non-state actors could benefit from fusion power plants for a nuclear weapons program to demonstrate real-world scenarios in which this may occur. The paper then gives a summary of current policy efforts in fusion power generation and highlights future topics may that require consideration, especially as the field expands in sophistication. Based on this review, it is apparent that fusion safeguards and nonproliferation safety is not currently progressing alongside developments in technology. A deliberate effort should be made to either include fusion under existing nuclear safeguard frameworks or strengthen existing export controls to maintain the peaceful use of fusion technology.

---

1   Captain Daniel Gum is a nuclear engineer for the Defense Threat Reduction Agency, currently stationed at Fort Belvoir, Virginia. The views expressed in this paper are his own and do not necessarily reflect the views of his employer, the Department of Defense, or the U.S. government.

INTRODUCTION TO NUCLEAR ENERGY

About 10 percent of the world's power comes from nuclear energy: a zero-carbon energy source with stable operation and consistent output of electricity to the grid.[2] This generation of energy functions on the principle of nuclear fission, whereby a heavy element (commonly uranium-235, or U-235) is broken in two and releases energy in the process. This form of energy generation has its drawbacks. For one, the waste from fission power is radioactive for long periods of time, meaning that the materials throughout the entire nuclear fuel cycle must be carefully controlled for safety, with the objective of avoiding negative impacts to human health and the environment. In addition, fission power plants can be used to create material for nuclear weapons, so they must be monitored and subject to rules and regulations that prevent the uncontrolled proliferation of such technology. Thankfully, there are robust and well-designed international and domestic agencies, regulations, and guidelines in place that aim to enable the peaceful use of nuclear technology while nullifying attempts to derive nuclear weapons from these same processes.

FISSION AND FUSION NUCLEAR POWER

Besides fission, fusion provides another way to generate energy though nuclear technology. This occurs when two lightweight elements, usually two hydrogen atoms, are combined to form a heavier element, such as helium, and in the process release energy that can be converted to electricity for use on the power grid. Fusion has long been seen as a promising new energy source that avoids many of the pitfalls of fission power, as the materials used for fuel are commonly found and do not produce extraordinarily long-lived radioactive waste upon use at a power plant. Fusion also manages to avoid the risk, at first glance, of nuclear weapons proliferation, as no output uranium or plutonium is generated during nominal operating conditions. Fission and fusion reactor cores are quite different in their design and energy production mechanisms, but the follow-on processes to generate electricity overlap significantly.

IN PURSUIT OF CLEANER ENERGY

In a global effort to combat climate change, leaders at the United Nation's Climate Change Conference created the breakthrough Paris Agreement, which aims to limit temperature increases related to global climate change compared to pre-industrial levels.[3] A cornerstone of this effort is increasing the diversification of the global energy portfolio into renewables and other low-carbon methods of power generation. As scientific research and development increase the accessibility of these technologies on the global market, newfound technologies have emerged as promising alternatives to oil and gas. Because of fusion technology's nascent state, fusion power is not seen as an alternative to fission or other low-carbon sources of energy in the goal to limit greenhouse gas emissions in the short term. Instead, it may be well suited to compliment low-carbon energy sources in providing power for the rapidly growing worldwide demand for energy in the latter half of this century.[4]

---

2   "Nuclear Explained," U.S. Energy Information Administration (EIA), 2016, https://www.eia.gov/energyexplained/nuclear/data-and-statistics.php.

3   "The Paris Agreement," United Nations Framework Convention on Climate Change, 2015, https://unfccc.int/process-and-meetings/the-paris-agreement/the-paris-agreement.

4   Robert J. Goldston and Jacob A Schwartz, "Fusion's Role in Fighting Climate Change," Bulletin of the Atomic Scientists, November 15, 2021, https://thebulletin.org/premium/2021-11/fusions-role-in-fighting-climate-change/.

Over the last decade, fusion energy has begun to emerge as a promising new energy source, with private companies raising over $1.9 billion in publicly declared funding, along with an additional $85 million in grants and other funding from governments as of 2021.[5] Among green energy enthusiasts, fusion energy for many decades was referred to as perpetually 20 years away, but now the outlook is changing. Many scientists now believe, however ambitiously, that fusion will be linked to the grid and delivering power by the 2030s.

There is a multitude of different approaches to this technology, with a large variety of fuel sources and nuclear reaction pathways. As a result, private companies are pursuing varied designs and even challenging long-favored preexisting routes to fusion in the process.[6] If these different approaches become commercially viable, there may be several competing designs and methods of fusion power generation. This necessitates a regulatory and safety framework specific to fusion to ensure proper operation of the wide range of these devices without inhibiting rapid innovation in the field.

Several organizations have begun exploring options for fusion regulation, including the International Atomic Energy Agency (IAEA) on the global scale and the Nuclear Regulatory Commission (NRC) and the UK Atomic Energy Authority (UKAEA) on a national level. However, thus far, no comprehensive set of rules and regulations exist to govern this emerging field. A few scientists have studied the issue in depth, but no broader efforts exist in the nuclear community to address the risks, as illustrated by Giorgio Franceschini et al.:

> An aspect often overlooked in the technology assessment and the theoretical debate on nuclear proliferation is the possible military dimension of fusion power.... Moreover, little attention has been given to the political conditions enabling (or constraining) such an option. As a consequence, there is practically no ongoing debate concerning the technological and regulatory options to prevent nuclear fusion from being used in a non-peaceful context.[7]

These devices need careful consideration, as they create a very different proliferation signature than a traditional fission plant, a trait which begins with the reaction itself and carries on to the methods of power generation used.

FUSION PROCESSES

The most commonly used fuel sources for a fusion reactor are deuterium and tritium, known as D-T fusion, but several other feasible fuel combinations exist as well: deuterium-deuterium (D-D), deuterium-helium-3 (D-HE3), and even hydrogen-boron fuels (p-B11).[8] Using these for a fusion reaction creates a wide range of outputs, neutron energy levels, spectra of emitted photons, and activated metals depending on which fuel source is selected. This diversity in fuel is matched by the variety of fusion systems being pursued. Magnetic confinement, inertial confinement, and magneto-inertial confinement are all competing approaches to power generation which use radically different

5   Melanie Windridge, Andrew Holland, and Tim Bestwick, *The Global Fusion Industry in 2021* (Washington, DC: The Fusion Industry Association and the UK Atomic Energy Authority, 2021).

6   Tom Clynes, "5 Big Ideas for Making Fusion Power a Reality," IEEE Spectrum, January 28, 2020, https://spectrum.ieee.org/5-big-ideas-for-making-fusion-power-a-reality.

7   Giorgio Franceschini, Matthias Englert, and Wolfgang Liebert, "Nuclear Fusion Power for Weapons Purposes," *The Nonproliferation Review* 20, no. 3 (2013): 525–44, doi:10.1080/10736700.2013.852876.

8   Fusion Industry Association, *The Global Fusion Industry in 2022* (Washington, DC: Fusion Industry Association, 2022), https://www.fusionindustryassociation.org/about-fusion-industry.

engineering architectures to achieve their goals. Because of this wide range of possible configurations for fusion power plants, an in-depth study of each individual system's proliferation resistance is required to adequately assess the risk that these devices pose to the global nonproliferation effort. In the meantime, it is possible to address the general proliferation risks that these devices present through analysis of the materials used and the outputs created during normal operations.

A general approach to safety and regulation is needed for these devices due to the usage of tritium, the activated materials in the reactor vessel, and the resulting neutron flux during operation. Of these concerns, the neutron flux is potentially the most concerning and most easily applicable to a nuclear weapons program. The commonly used D-T reaction produces an energetic neutron which can be used to transmute nuclear material via conversion of uranium-238 (U-238) into the fissile, weapons-quality plutonium-239 (Pu-239), or be used for production of uranium-233 (U-233) via irradiation of thorium-232 (Th-232).[9]

The proliferation risk is linked to the production of electricity in many of these systems. The resulting nuclear reactions are contained in a steel vessel which is lined with a fusion "blanket" which allows for conversion of neutronic energy into heat and, later, electricity.[10] A would-be proliferator, here defined as a state or non-state actor working to obtain a nuclear weapon, could seek to line this blanket with the aforementioned materials U-238 and Th-232 to absorb some of the neutron flux and convert fertile material into fissile material. This strategy would be difficult to pursue openly, as a commercial fusion reactor does not require any fertile elements for operation, and thus any introduction of U-238 or Th-232 into a fusion facility would be immediately noticed and flagged as unusual activity requiring investigation by the international safeguards community. There are still some upsides to this approach that might attract nuclear weapons-seeking actors. The amount of initial natural—or even depleted—uranium needed to produce a nuclear weapon could be achieved with as little as several hundred kilograms of uranium; a fission reactor attempting to do the same would require around 10 metric tons of natural or enriched uranium to operate the reactor and develop ample nuclear weapons material.[11]

ADDITIONAL PROLIFERATION CONCERNS

Another aspect to consider with fusion reactors is that they can produce tritium through normal reactor operations, a material with loose international safeguards which can be used to boost the yield of nuclear weapons.[12] This production of tritium would be necessary to create a self-sustaining fuel cycle when using D-T fuel, as the global supply of tritium is extremely limited. The ambiguity over creation of nuclear weapons material rather than valid fusion fuel production makes for a tricky environment to properly regulate. It could be argued that the main proliferation concern is that reactor blankets could be modified for use completely outside of normal operating conditions through the insertion of fertile material with the explicit goal of fissile material production for

9   T.C. Simonen, R.W. Moir, A.W. Molvik, and D.D. Ryutov, "A 14 MeV fusion neutron source for material and blanket development and fission fuel production," *Nuclear Fusion* 53, no. 6 (April 2013), doi:10.1088/0029-5515/53/6/063002.

10 K. Hesch, L.V. Boccaccini, and R. Stieglitz, "Blankets – key element of a fusion reactor – functions, design and present state of development," *Kerntechnik* 83, no. 3 (June 2018), 241–250, doi:10.3139/124.110923.

11 Matthias Englery, Giorgio Franceschini, and Wolfgang Liebert, "Strong Neutron Sources - How to Cope with Weapon Material Production Capabilities of Fusion and Spallation Neutron Sources?," 7th INMM/Esarda Workshop, Aix-en-Provence, 2011, https://www.hsfk.de/fileadmin/HSFK/hsfk_downloads/Strong_Neutron_Sources_final.pdf.

12 Robert Kelley, "Starve Nuclear Weapons to Death with a Tritium Freeze," Stockholm International Peace Research Institute, August 28, 2020, https://www.sipri.org/commentary/topical-backgrounder/2020/starve-nuclear-weapons-death-tritium-freeze.

nuclear weapons purposes. However, this is not the only issue at hand: a fusion reactor contains an environment with a variety of proliferation concerns, some more minor than others, but all requiring at least some degree of consideration.

If either a fusion or a fission power plant were used to create fissile material, the quantities involved would need to approach a certain threshold before an intended nuclear weapons program could no longer be ruled out. The IAEA defines an amount of material from elements deemed significant to nuclear weapons production as a "significant quantity" (SQ) when "the possibility of manufacturing a nuclear explosive device cannot be excluded."[13] These quantities include 8 kg of Pu-239 or 8 kg of U-233. Ideally, international safeguards or strict export controls would be put into place to prevent the production of SQs of nuclear material through use of any variety of nuclear power plant. To prevent the spread of nuclear weapons from these power plants, there are three different scenarios of differing levels of feasibility that would need to be considered: clandestine production in an undeclared facility, covert production of material in a declared facility, and the overt use of a declared facility in a breakout scenario. These three scenarios are discussed below using magnetic confinement fusion approaches, as this type of fusion is currently regarded as the most mature technology. A follow-on discussion covers the other two main types of fusion—inertial and magneto-inertial confinement—which are rapidly catching up to magnetic confinement efforts and are seen as promising alternatives in the path to commercial fusion power.

USING A FUSION PLANT TO MAKE FISSILE MATERIAL: A CLANDESTINE APPROACH

Clandestine production of nuclear weapons has long been a concern of the international community, with countries such as Iran serving as textbook examples of programs that remained covert while attempting to build an arsenal of nuclear weapons. Completely hiding a fusion power plant from the public eye has been assessed as highly unlikely due to a variety of factors. Primarily, the size of a fusion plant itself makes these systems difficult to hide: on a large scale, the ITER reactor Tokamak measures nearly 30 meters in height, while a mid-scale reactor such as Commonwealth Fusion System's SPARC stands at 12 meters in height. These fusion reactors also require extensive power, cooling, and control systems which dramatically increase the overall footprint of fusion power plants: a mid-scale reactor such as the Tokamak Fusion Test Reactor (TFTR) sits on a 10-hectare engineering campus, with a building footprint of more than 7,000 square meters, non-inclusive of the required power substation, control room, or cooling tower.

Reactors currently in use and those in construction for power generation in the near future all deliver fusion power on the megawatt (MW) scale. These facilities also require MW-scale input power to operate and have a hypothetical fissile material production capability of roughly 3.5 kg of Pu-239 or U-233 per year, not even a significant quantity.[14] Even downsizing the reactor in an attempt to reduce the physical footprint produces a facility that is still large enough to have its own distinct environmental signatures: the use of tritium in fusion reactors can be distinguished from background activity, with prominent reactors such as the TFTR reporting trace levels of tritium detectable within tens of kilometers from the

13   International Atomic Energy Agency, *IAEA Safeguards Glossary* (Vienna: 2001), https://digitallibrary.un.org/record/475621?l-n=en.

14   Robert J. Goldston, Alexander Glaser, and A.F. Ross, "Proliferation Risks of Magnetic Fusion Energy: Clandestine Production, Covert Production and Breakout," *Nuclear Fusion* 52, no. 4 (2012): 043004, doi:10.1088/0029-5515/52/4/043004.

facility.[15] Overall, fusion power plants contain many unique signatures during operation that would allow for detection of proliferant activates and as such do not pose a significant proliferation concern under clandestine conditions with current and near-term technology levels.

USING A FUSION PLANT TO MAKE FISSILE MATERIAL THROUGH COVERT PRODUCTION

Covert production of nuclear weapons material would involve altering the usage of a publicly declared fusion plant to generate fissile material without alerting the international community. The most likely scenario would involve modification of the fusion blanket through the addition of small amounts of fertile material so that it may absorb neutrons over long timescales. In practice, this requires diversion of U-238 or Th-232 from clandestine facilities or existing safeguarded stockpiles with the intent of making fissile material while avoiding detection. Proper detection and prevention methods here would need to be minimally invasive to the overall fusion power plant architecture to prevent interference in normal power plant processes and to place a minimal burden on power plant operators. One method of doing so would be the detection of specific gamma ray emissions from the fertile or fissile material in the fusion blanket structure. In the case of transmutation from uranium to plutonium, detectors could search for the characteristic gamma ray line emitted by U-238.[16] For the production of weapons-grade uranium, detectors could search for emissions from natural thorium or the resulting U-233, although this is more difficult.[17] In both cases, detection scenarios appear to be clearly defined, but more studies are required on the background radiation levels expected at fusion power plants operateing on the MW scale or larger to properly measure material quantities that could be detected, as well as measure time to detection. In the case of liquid reactor coolant mixed with fertile materials, inspectors could also look for undeclared injection and extraction systems in addition to random testing of the coolant to check for fertile and fissile elements.

For fusion blanket modules that are solid as opposed to liquid metal, the proliferation process becomes more involved and requires addition of fertile materials to the fusion blanket during initial fabrication of the blanket. Detection of illicit materials would require inspection of new components added to the fusion reactor either through the gamma and neutron detection methods outlined above or with active interrogation through use of portable neutron generators.[18] The lack of fertile or fissile material on site during routine fusion power plant operations is the key for detection of any covert weapons material generation. These additional inspection techniques are relatively easy and noninvasive and provide strong assurances of safe and secure nuclear power use.

<hr>

15   Ibid.

16   Ibid.

17   Jonathan Medalia, *Detection of Nuclear Weapons and Materials: Science, Technologies, Observations*, CRS Report No. R40154 (Washington, DC: Congressional Research Service, August 2009), https://sgp.fas.org/crs/nuke/R40154.pdf; and Louise Evans-Worrall, "Safeguards Technology Considerations and Research Needs for Thorium Fuel Cycles and Molten Salt Reactors," (presentation, National Academies of Sciences, October 15, 2021), https://www.nationalacademies.org/documents/embed/link/LF2255DA3DD1C41C0A42D3BEF0989ACAECE3053A6A9B/file/D758372FB4913EF59E0BF6BBA-926DE1219C8277A72D9?noSaveAs=1.

18   David Chichester and Edward Seabury, "Active Neutron Interrogation to Detect Shielded Fissionable Material," Idaho National Laboratory, International Topical Meeting on Nuclear Research Applications and Utilization of Accelerators, May 2009, https://inldigitallibrary.inl.gov/sites/sti/sti/4480305.pdf.

USING A FUSION PLANT TO MAKE FISSILE MATERIAL IN A BREAKOUT SCENARIO

In a breakout scenario, a nation operating a fusion power plant under IAEA or other international regulations expels inspectors and begins producing nuclear weapons material as quickly as possible in a race against international sanctions and intervention. Such a scenario would require extensive preparation before the breakout: large amounts of nuclear material would need to be collected and properly manufactured to fit within liquid or solid metal blankets. In contrast to a fission breakout scenario, a fusion breakout contains no fertile nuclear material in the power plant to begin with, so would-be proliferators would need to acquire these materials through other means such as purchase of natural uranium or natural thorium, or diversion of these materials from traditional fission power plants. Even after obtaining these materials, introducing them into a fusion blanket and properly installing the liner is a nontrivial process, currently estimated at one month in duration for a liquid system and longer for a solid fusion blanket.[19] Once operational in a breakout scenario, analyses have shown that a fusion power plant could produce fissile material at a rate comparable to a fast-spectrum fission breeder reactor of similar power output levels.[20] However, these quick production rates may not be sustainable, as the power required could put the tritium replenishment levels below what is needed to maintain a stockpile, and thus the nation-state or non-state actor will end up limited by their tritium supply in addition to their finite fertile material.

In summary, a timescale of low single-digit months would be required to produce one SQ of nuclear weapons material in a fusion power plant. This scenario can be prevented through the usual political, diplomatic, and economic actions, but unlike fission power plants, there are several more overt courses of action that can be taken. Much of the footprint of fusion power plants consists of supporting systems that are non-nuclear in nature. Because fusion plants do not rely on a metastable chain reaction for operation and are therefore fail-safe, these supporting systems may be disabled or entirely removed to safely shut down suspected weapons-producing activities before they can yield fissile material. These vulnerable systems include the power conditioning equipment used to run the large magnets needed for fusion, as well as the cooling plants and water management systems. Disabling these systems through military action, cyber warfare actions, or other methods would allow for a minimized risk of nuclear contamination while completely halting power plant operations.

INERTIAL CONFINEMENT FUSION

Fusion research began from two main approaches to obtaining net energy gain from nuclear reactions: magnetic confinement fusion and inertial confinement fusion (ICF). This second type consists of a process which rapidly compresses small containers filled with thermonuclear fuel, usually deuterium and tritium, to achieve nuclear fusion. This compression is accomplished through the use of powerful laser arrays that compress the material and ignite the fusion process. The most well-known ICF experiment is the National Ignition Facility (NIF) at Lawrence Livermore National Laboratory (LLNL), which serves a multipurpose role beyond its exploration of fusion science. As the LLNL explains, "the drive toward ignition helps ensure the reliability of the nuclear deterrent, while also opening new frontiers in laboratory astrophysics, materials science, hydrodynamics, and many other scientific disciplines."[21] The NIF experiment accomplishes many different goals in basic and

---

19   Glaser and Goldston, "Proliferation Risks of Magnetic Fusion Energy."
20   Ibid.
21   "What Is NIF?," Lawrence Livermore National Laboratory, https://lasers.llnl.gov/about/what-is-nif.

applied science, but its main mission is that of stockpile stewardship and the study of matter under the extreme conditions found in nuclear weapons detonations. This brings forth another aspect of nuclear proliferation, whereby nation-states interested in furthering their understanding of nuclear weapons may seek out ICF systems as a means to an end. Put plainly, "the risk of broad dissemination of scientific information … [helping] nations design very compact, light, and powerful nuclear weapons that can be delivered by low throw-weight, long-range missiles" could be a very real risk with uncontrolled dissemination of ICF systems and the knowledge gained from their operation.[22]

This type of knowledge is particularly valuable in the face of restricted nuclear weapons testing agreements such as the Comprehensive Nuclear-Test Ban Treaty. After the initial quote of $2.1 billion for NIF construction, the U.S. government would go on to spend a total of $3.9 billion to finish the project, thus putting a direct price tag on the study of thermonuclear weapons environments in support of stockpile stewardship.[23] The basic concepts behind NIF and other ICF tests were closely guarded and considered classified by the United States until 1979.[24] In 1995, the U.S. government undertook a review of the proliferation risks associated with ICF systems and noted that an ICF program would allow for an advanced proliferator to keep a knowledgeable network of scientists and engineers under the pretense of a legitimate, peaceful nuclear activity.[25] The 1995 report goes on to state that "Without nuclear testing, it is probable that a proliferator would not be able to develop a highly deliverable thermonuclear weapon," suggesting that ICF testing would be able to lessen the quantity of nuclear detonation tests required through use of the supplemental information gained. Thankfully, the large cost of these systems greatly lessens the overall proliferation risk, but there are a few companies using this sort of technology as their approach to creating fusion reactions that may require careful regulation as their science matures to commercial scale.

COMMERCIAL ICF APPROACHES

Of the 23 companies that responded to the Fusion Industry Association's comprehensive 2021 survey, only two reported seeking a standard ICF approach to producing net energy gain. The first is Marvel Fusion, a company started in 2019 in Germany which aims for a pilot plant in the 2030s. This approach focuses on a p-B11 reaction, which has several advantages in terms of proliferation resistance in that it is aneutronic and, thus, initially seems to avoid the strong neutron source problem seen in more traditional fusion reaction schemes using D-T fuels.[26] However, it is commonly understood in the fusion industry that there are no true aneutronic reactions, only slightly less neutronic reactions. This is the case with p-B11 fuel, which may not produce the prodigious number of neutrons seen in D-T fuel sources but still creates a sufficiently large amount to require proper shielding to protect against adverse human effects.[27] These "aneutronic" fuels require further study

22   Robert J. Goldston and Alexander Glaser, "Inertial Confinement Fusion Energy R&D and Nuclear Proliferation: The Need for Direct and Transparent Review," *Bulletin of the Atomic Scientists* 67, no. 3 (2011): 59–66, doi:10.1177/0096340211407562.

23   Government Accountability Office, *Department of Energy: Follow-up Review of the National Ignition Facility*, GAO-01-677R (Washington, DC: 2001), https://www.gao.gov/products/gao-01-677r.

24   Office of Health, Safety and Security, Office of Classification, *Restricted Data Declassification Decisions, 1946 to the Present (RDD-8)* (Washington, DC: U.S. Energy Department, 2002), https://sgp.fas.org/othergov/doe/rdd-8.pdf.

25   Office of Nonproliferation and National Security, "Study on the National Ignition Facility and the Issue of Nonproliferation," *Federal Register* 60, no. 45150 (1995), https://www.govinfo.gov/app/details/FR-1995-08-30/95-21536.

26   A.A. Harms et al., Principles of Fusion Energy: An Introduction to Fusion Energy for Students of Science and Engineering (River Edge, NJ: World Scientific, 2000).

27   M. Heindler and W. Kernbichler "Advanced Fuel Fusion," in *Proceedings of the 5th International Conference on Emerging Nuclear Energy Systems*, edited by Ulrich von Möllendorfd and Balbir Goel (Singapore: World Scientific, 1989): 177–82.

to properly characterize their ability to convert fertile material to fissile weapons material in a timeframe useful to state or non-state actors seeking to acquire a nuclear weapon. One safeguards benefit of many of these aneutronic fuels is that they do not require tritium breeding, which alleviates some of the secondary proliferation concerns seen with other fusion methods by avoiding the need to keep a large tritium stockpile on hand.

The second ICF company polled was First Light Fusion, a company spun out of Oxford University in 2011 which aims to construct a pilot plant in the 2030s. First Light uses a unique approach to ICF, whereby a solid projectile initiates the fusion reaction instead of a laser pulse as in most other modern ICF approaches.[28] This may help avoid the sensitivities surrounding the decades of laser-based ICF, but their D-T reaction method combined with lithium coolant walls raises the same questions as magnetic confinement reactors in regard to the viability of proliferant activities. In short, ICF methods may not currently be the most popular approach to fusion power generation, but it is important to note that their operation does overlap with many tests and experiments used to refine nuclear weapons arsenals. The majority of these ICF companies aim to have viable pilot plants in the 2030s and will need a closer look by regulating authorities to ensure a proper, peaceful use of their technologies.

MAGNETO-INERTIAL CONFINEMENT

Several private companies are taking a third route to fusion by choosing a hybrid approach of inertial confinement and magnetic confinement, known as magneto-inertial confinement. Of the companies polled by the Fusion Industry Association, this variation of fusion power generation is the second most popular. Much like ICF and magnetic confinement, there are a wide variety of techniques involved, which complicates addressing the field with generic proliferation concerns and solutions. Two up-and-coming private companies, TAE Technologies and Helion Energy, both use field-reversed configuration (FRC) devices to create electricity, unique in their approach to clean power generation.[29] Both of these companies are pursuing aneutronic fusion, with Helion producing D-He3 reactions and TAE Technologies using p-B11 for fusion fuel. As seen previously, these reactions are commonly labeled as aneutronic but still actually produce significant quantities of energetic neutrons that could feasibly be used to convert fertile material into fissile weapons material.[30] Other companies in the inertial-magnetic design space, such as Generalfusion, use the more common D-T fuel but also use unique reactor architectures that have not been previously studied under the lens of nuclear proliferation.[31] A further look by the IAEA and national-level regulators is warranted to maintain the growth of fusion safeguards in tandem with the technological advancement in this type of nuclear power plant.

POTENTIAL USE CASES

As scientific investments in fusion technology increase over time, the cost of joining the fusion community of interest will decrease, but likely not enough to make such capabilities viable for any but the most technologically advanced nations in the near term. Once commercially viable, fusion would probably remain within a select few industrialized states, mostly likely those with significant backgrounds in the technology itself. Currently under construction, the ITER reactor is an international collaboration and a flagship of the fusion community run by seven members: China, the European

28   James Varley, "First Light Gives It Its Best Shot," Nuclear Engineering International, March 25, 2021, https://www.neimaga-
      zine.com/features/featurefirst-light-gives-it-its-best-shot-8625980/.
29   "Frequently Asked Questions," TAE Technologies, March 18, 2014, https://tae.com/faq/.
30   Heindler and Kernbichler, "Advanced Fuel Fusion."
31   "Our Technology - Practical Fusion Power Technology," General Fusion, n.d., https://generalfusion.com/fusion-technology/.

Union, India, Japan, South Korea, Russia, and the United States. Many within this group have extensive experience with fission power and thus have traditional proliferation routes open to them, but countries within the European Union—Italy, which does not have a nuclear program, or even Germany and Switzerland, who are currently phasing out their fission power plants—may seek to re-enter the nuclear community through fusion technology.[32] This trend can even hold for countries that have fission power plants: due to rising tensions in their local geopolitical regions, Taiwan, South Korea, Japan, and India may seek fusion technology to provide them a new and unregulated proliferation capability.[33] On the other hand, the potential for illicit use of these power plants by a country with low levels of technological sophistication remains small in the near future. However, given a long enough timeframe, fusion technology can provide non-nuclear states the capability to produce a homegrown nuclear weapons program. This further reinforces the need to create specific safeguards and export controls to address the peaceful uses of this technology for nuclear and non-nuclear states alike.

The majority of the discussion around the nuclear weapons potential of fusion technology centers around the creation of traditional uranium or plutonium weapons. It is worth briefly mentioning an alternative weapons use of fusion power plants: dirty bombs. Despite being consistently touted as a source of clean energy that avoids much of the radioactive waste issues from fission power, fusion still creates intensely radioactive waste due to the highly neutronic environments that many fusion fuels create. The container vessels that surround the fusion plasma are constantly bombarded by energetic neutrons, which activates the container materials and makes them radioactive. The neutron flux inside a fusion power plant is many orders of magnitude higher than in a fission power plant, which makes the reactor walls more radioactive compared to those in a fission plant. If a malevolent actor so chose, these materials could be used as a dirty bomb: a study by David Petti et al. found that a sudden release of 150 grams of tritium or of 6 kilograms of reactor vessel dust would require evacuation from the facility and the surrounding areas.[34] If so desired, these materials could be dispersed from an explosion on site or packaged into a bomb designed to spread contamination elsewhere. It is unlikely, however, that tritium would be chosen as the radioactive material for this hypothetical dirty bomb, as the cost on the open market approaches $30,000 per gram.[35]

When combined with the enormous cost and complexity required to build a fusion plant, it becomes unlikely that either tritium or reactor vessel dust would willingly be used to create dirty bombs when easier routes to the same end goal exist within other already established means. The Andlinger Center summarizes nicely:

> Overall, the fusion power system presents far smaller risks than the fission power system from the point of view of becoming associated with the malevolent dispersal of radioactivity. Neither the risk of an attack on fusion power infrastructure nor the risk of using fusion waste in a dirty bomb would seem to be significant, though the risk is present. By contrast, despite the numerous safety measures in place, the comparable risks from the fission power system are far higher, both because of the possibility of meltdown and because of the much greater quantities of highly radioactive materials involved.[36]

32   Franceschini, Englert, and Liebert, "Nuclear Fusion Power for Weapons Purposes."
33   Ibid.
34   D.A. Petti et al., "ARIES-at Safety Design and Analysis," *Fusion Engineering and Design* 80, no. 1 (2006): 111–37, doi:10.1016/j.fusengdes.2005.06.351.
35   Jane Baldwin et al., "Fusion Energy via Magnetic Confinement," Andlinger Center for Energy and the Environment, Princeton University, April 2016, 2022, https://acee.princeton.edu/distillates/fusion-energy-via-magnetic-confinement/#technology.
36   Ibid.

POLICY IN FUSION POWER

The creation of fission technology was inherently tied to nuclear weapons, and the duration of World War II was marked by a highly sophisticated arms race to develop the first atomic bomb. Only after the war did the peaceful uses of nuclear power gain their full relevance in society, with the technology to generate power rapidly spreading around the world: in the span of 45 years, 31 countries had developed, or were developing, their own nuclear power plants. Because of the recency of the bomb and its unprecedented destruction, a particular emphasis was placed on the peaceful aspects of nuclear technology, with organizations such as the Atomic Energy Commission and UKAEA forming soon after the war to oversee the development of nuclear technology. National-level organizations were followed by the establishment of the IAEA in 1957 to coordinate the peaceful use of nuclear technology on a global scale. Follow-on international treaties such as the Limited Test Ban Treaty and the Treaty on the Nonproliferation of Nuclear Weapons limited the testing, manufacturing, and development of nuclear weapons in exchange for greater scientific cooperation and nuclear technology sharing.[37] In the decades following the advent of nuclear weapons, policy and technology progressed steadily in tandem.

Perhaps because the most pressing needs of global nuclear nonproliferation lie in topics immediately at hand, or because of the long-unfulfilled promise of fusion technology, little concerted effort has been made to guide the technology from a regulatory standpoint. This may be a good thing for now, as there are no burdensome requirements that could stifle creativity and the current atmosphere of accelerated technological development. But the field should not be left on its own, as many emerging private companies are aiming for large scale pilot plants in the 2030s for viable demonstrations of net energy gain from fusion reactions, with the intent of manufacturing power plants for widespread commercial use thereafter. Thankfully, the issue has started to gain traction among national governing bodies aware of the rapid pace of technological development and keen to fill the regulatory vacuum. The European Union has begun to explore regulatory options regarding fusion power plants, with a particular emphasis on the safety of systems against events that may be hazardous to human health.[38] The European Union undertook a review of these power plants, finding that "presently no country has a dedicated and specific regulatory framework for fusion facilities. Nevertheless, it was observed that the competent regulatory authorities in more and more countries pay attention to this issue."[39] This is a good start, and it is reassuring to see that there has been a growing recognition of the need for an integrated approach to nuclear safety, security, and safeguards.

DOMESTIC POLICY APPROACHES

In 2018, the U.S. Congress passed the Nuclear Energy Innovation and Modernization Act (NEIMA), which required the NRC to address the challenges involved in licensing advanced nuclear reactors, a category which includes both fission and fusion.[40] To accomplish this goal, the NRC has reached out

37   "Treaty Banning Nuclear Weapon Tests in the Atmosphere, in Outer Space, and Under Water," U.S. Department of State, https://2009-2017.state.gov/t/avc/trty/199116.htm; and "Treaty on the Non-Proliferation of Nuclear Weapons (NPT)," United Nations, n.d., https://www.un.org/disarmament/wmd/nuclear/npt/.

38   Lars-Goran Erikkson et al., *Exploring Regulatory Options for Fusion Power Plants* (Brussels: European Commission, 2021), doi:10.2777/980320.

39   Directorate-General for Energy, Study on the applicability of the regulatory framework for nuclear facilities to fusion facilities: towards a specific regulatory framework for fusion facilities: final report (Brussels: European Commission, 2022), doi:10.2833/787609.

40   R. Patrick White and Liam S. Hines, "Do proposed regulatory approaches for commercial fusion energy jeopardize successful technology deployment?," *MIT Science Policy Review* 2 (2021): 40–45, doi:10.38105/zv7f0d3vxd.

to stakeholders and industry partners in an attempt to involve the larger community of interest from the very beginning. In 2021, the NRC started a workshop series to discuss the variety of regulatory frameworks that might be implemented in the fusion power industry. No final decision has been made, but ongoing discussions have centered on the need to properly regulate the industry as a distinct entity from fission power while also allowing for rapid growth and innovation. The NRC is currently evaluating a variety of regulatory models that would enable this state of operations, but once again the discussion is almost entirely focused on physical safety over proliferation safety. In the continued exchanges that take place between nuclear states and their stakeholders, a safeguards-by-design approach must be heavily considered, especially as demonstration power plants scale into full capacity in the coming decades. This can only be accomplished by continual cooperation between governing bodies and fast-moving private fusion companies.

In addition to the regulations created by the IAEA, nuclear technology is protected via the use of export controls that are specific to each individual country. These export controls serve to prevent states and non-state actors from acquiring materials or technology that could lead to the production of nuclear weapons. An overarching regulatory framework is provided by the IAEA, but final authority for controlling international trade in nuclear technology lies with national level regulatory bodies through an exception to the General Agreement on Tariffs and Trade.[41] To prepare for the future of the fusion industry, these domestic processes require investment to bring them in line with the requirements of nuclear technology that does not fall within the traditional framework of fission power plants. As noted by the White House, "we need to begin work on harmonizing international licensing standards and permitting processes to support export markets, including building in the appropriate safeguards and non-proliferation measures to underpin export control regulations."[42]

Fusion power plants represent groundbreaking, state-of-the-art engineering projects that will demonstrate new technology at an industrial scale. This will bring new technology to the forefront of power generation, and many systems will not have the benefit of decades of trial and error on which to make well-informed regulatory decisions. Novel technologies can be operated in a safe and consistent manner, but this development is an iterative process that will require a continual back and forth between policymakers, engineers, and scientists. National-level regulatory bodies must partner with international organizations such as the IAEA to properly address the wide variety of approaches to fusion power and adequately disassociate fusion regulations from a fission-informed approach to safeguards and controls.

CONCLUSION

A review of the current status of private fusion research and development efforts shows that there is a myriad of approaches to producing net energy gain, but most can be broken down into three categories: magnetic, inertial, and magneto-inertial confinement. Of the three, magnetic is the most developed, but inertial and magneto-inertial have produced promising results due to recent advances in technology and should be considered viable alternatives or even possible front-runners in the race to commercialized fusion. The proliferation risks from such devices vary among the different approaches. Several authors in the open literature have examined general trends in

41   World Nuclear Association, *An Effective Export Control Regime for a Global Industry* (London: April 2018), https://www.world-nuclear.org/getmedia/cc6d54da-ee87-4642-aee3-99e0231016d9/Export-Controls-Report.pdf.aspx.

42   Sally M. Benson and Costa Samaras, "Parallel Processing the Path to Commercialization of Fusion Energy," The White House, June 6, 2022, https://www.whitehouse.gov/ostp/news-updates/2022/06/03/parallel-processing-the-path-to-commercialization-of-fusion-energy/.

nuclear proliferation from the perspective of both magnetic and inertial confinement systems, but no widespread effort has been made to address specific systems and their unique proliferation resistances (or lack thereof), unlike in fission power plants where this has been extensively studied. Much of the reason for this is the cutting-edge nature of many of these different approaches to fusion: a large number of them are currently publishing new updates to their methodology in prominent science journals as their techniques improve and progress in sophistication.

Regardless, to ensure safe and proper use of these systems, a comprehensive look at each and every variety of fusion technique is warranted. This should be done as soon as possible to aid design of efficient and minimally burdensome safeguards and controls, as that will yield the greatest chance of success in deterring illicit activities. To take a cue from the world of fission power engineering, new fusion power plants must aim for an approach of "safeguards-by-design" to ensure the safest operating conditions possible for the broader global community. In addition, existing IAEA safeguards and glossaries must be updated to include fusion technologies in their regulatory frameworks in a further effort to cleanly separate fusion technology and materials from fission technology. Fusion can be done in a manner that is both safe and secure, but it is a specific and measured decision that must be made to prepare for technology that will soon be in use and is closely emerging on the horizon.

Chinese War Controls and U.S. Approaches to Deterrence

By Annika Kastetter[1]

War control, a central Chinese strategic concept, is gaining attention among Western military analysts.[2] According to authoritative People's Liberation Army (PLA) sources, war control involves employing various elements of national power to precisely control conflict at all levels and phases to achieve key political objectives at the smallest possible cost.[3] The notion that war can be precisely steered to serve established political ends diverges from Western military thinking, which emphasizes the inherent uncertainty and "fog of war."[4] Indeed, Western scholars and strategists largely contend that escalation and conflict can be managed but not controlled.[5]

War control has significant implications for U.S. deterrence planning and approaches. Specifically, if Chinese leaders believe they can control the timing, scale, scope, and intensity of operations across the spectrum of conflict, including an adversary's reactions, they may advertently or inadvertently undertake provocative measures that raise the risk of escalation. Given the stakes of an armed

---

1   Annika Kastetter is a defense strategy and policy analyst at Systems Planning and Analysis, Inc. The views expressed in this paper are her own and do not necessarily reflect the views of her current employer or clients.

2   The Office of the Secretary of Defense's annual report to Congress, *Military and Security Developments Involving the People's Republic of China*, uses the term "effective control" to define the Chinese war control concept. The 2013 and 2020 editions of the *Science of Military Strategy* reference "effective control," "war control," and "war situation control."

3   Burgess Laird, *Managing Escalation and Limiting War to Achieve National Objectives in a Conflict in the Western Pacific* (Washington, DC: Center for New American Security, August 2016), 4, https://s3.us-east-1.amazonaws.com/files.cnas.org/documents/CNASReport-ChineseDescalation-Final.pdf; Lonnie D. Henley, "War Control," in *Shaping China's Security Environment: The Role of the People's Liberation Army*, edited by Andrew Scobell and Larry M. Wortzel (Carlisle, PA: Strategic Studies Institute, U.S. Army War College, 2005): 82–83, https://press.armywarcollege.edu/cgi/viewcontent.cgi?article=1709&context=monographs; and Burgess Laird, *War Control: Chinese Writings on the Control of Escalation in Crisis and Conflict* (Washington, DC: Center for New American Security, April 2017): 12, https://www.cnas.org/publications/reports/war-control.

4   Carl Von Clausewitz, *On War* (New Jersey: Princeton University Press, 1976), 85, 119–21, 140.

5   See Forrest Morgan et al., *Dangerous Thresholds: Managing Escalation in the 21st Century* (Santa Monica, CA: RAND Corporation, 2008): xii, 7–45, https://www.rand.org/pubs/monographs/MG614.html.

conflict between the United States and China, U.S. strategists should develop a tailored deterrence strategy that accounts for the Chinese war control concept.

This paper aims to contribute to this discussion by examining how China's war control concept affects U.S. deterrence planning and by proposing a framework for how the United States should think about and strengthen deterrence in the context of war control.

WAR CONTROL: A NOTE ON SOURCES

War control is an evolving concept that has become increasingly prominent in PLA writings over the last two decades. The term "war control" began appearing in official PLA sources in the early 2000s.[6] The first Western study on the topic was published in 2006 by Lonnie Henley, a former defense intelligence officer for East Asia.[7] More recent and extensive primary discussions of war control appear in the 2013 and 2020 editions of the *Science of Military Strategy* (SMS) published by the PLA's Academy of Military Science.[8] Importantly, the 2013 and 2020 SMS editions discuss war control within the broader context of "effective control," which appears to be an evolution of earlier thinking on the war control concept.[9]

War control and its associated concepts have received recent attention in the U.S. defense community. In its 2021 *Annual Report to Congress on Military and Security Developments Involving the People's Republic of China*, the Office of the Secretary of Defense devoted a section to "The PRC's Effective Control Concept and PLA Escalation Management Views."[10] In September 2022, the U.S. Department of the Air Force's China Aerospace Studies Institute (CASI) published a monograph called *Chinese Views of Effective Control: Theory and Action*.[11]

Although some secondary sources highlight deterrence and escalation challenges associated with war control, this paper distinguishes itself by examining how the United States should tailor its deterrence strategy in response to China's war control concept.

FRAMING WAR CONTROL

PLA writings on war control posit that various elements of national power can be leveraged to precisely control the timing, scale, scope, and intensity of operations across the spectrum of conflict, including an adversary's perceptions and reactions, to achieve defined political objectives.[12] The concept of war

---

6   Henley, "War Control," 82.

7   See ibid.

8   The *Science of Military Strategy* is a capstone document on China's current military strategy.

9   See Daniel Shatz, *Chinese Views of Effective Control: Theory and Action* (Montgomery, AL: China Aerospace Studies Institute, Air University, September 2022): 2, https://www.airuniversity.af.edu/Portals/10/CASI/documents/Research/Other-Topics/2022-10-03%20Effective%20Control.pdf. The report states: "Effective control is composed of the three core concepts of 'establishing posture' during peacetime (using generally nonviolent means to shore up China's strategic weaknesses relative to adversaries, maintain stability on China's periphery, and curb Taiwanese 'separatism'); preventing and controlling crises (with the dual goals of preventing escalation into war while also 'seizing opportunities' to advance Chinese interests); and controlling war situations when war does erupt (seeking to win as quickly as possible, at minimal cost, while containing escalation in intensity, scope, and other dimensions)."

10   Office of the Secretary of Defense, *Military and Security Developments Involving the People's Republic of China 2021: Annual Report to Congress* (Washington, DC: Department of Defense, November 2021): 155–56, https://media.defense.gov/2021/Nov/03/2002885874/-1/-1/0/2021-CMPR-FINAL.PDF.

11   See Shatz, *Chinese Views of Effective Control*.

12   China Aerospace Studies Institute, *PLA's Science of Military Strategy (2013)* (Maxwell Air Force Base, AL: Air University, February 2021): 137–42, https://www.airuniversity.af.edu/CASI/Display/Article/2485204/plas-science-of-military-strategy-2013/; China Aerospace Studies Institute, *In Their Own Words: 2020 Science of Military Strategy* (Maxwell Air Force Base,

control stems from the belief that crises and conflicts must be controlled to prevent economic turmoil and disruptions to national development during the period of "strategic opportunity."[13]

Beyond underscoring the strategic *need* to exercise war control, authoritative PLA sources contend that technological advancements, combined with meticulous planning, make it increasingly *possible* to control activities across the spectrum of conflict.[14] PLA writings emphasize that high-technology capabilities designed to operate "under conditions of informatization and intelligence"—or conditions where advanced information technologies, information-based capabilities, and strategic intelligence preparation provide unprecedented access to and use of information during a conflict—enable military commanders to precisely plan for, execute, and control warfare at all levels and across all domains.[15]

For instance, improvements in intelligence, surveillance, and reconnaissance (ISR), command and control (C2), and precision-guided weapons are believed to support war control by increasing battlefield transparency, perfecting situational awareness, strengthening operational control, and enabling highly accurate, massed strikes from extended ranges.[16] PLA writings also suggest that nonlethal attack methods, such as cyber, can limit wartime violence and enhance the PLA's ability to control the scale and intensity of a war.[17] Moreover, sophisticated information technologies combined with artificial intelligence (AI) are believed to help Chinese planners understand adversaries and anticipate adversary responses at the tactical, operational, and strategic levels.

In sum, PLA writings on war control largely frame modern warfare as a scientific, deterministic process that can be precisely engineered to control adversary behavior and achieve desired outcomes. The 2013 and 2020 editions of the SMS do acknowledge, however, that the "fog of war" as well as obstacles related to intelligence, information, and technology present challenges for war control and must be accounted for.[18]

KEY WAR CONTROL CONCEPTS

War control features two key concepts that are crucial for U.S. deterrence planning: "seizing the initiative" and "inexorable momentum." Advanced battlefield technologies, informatized conditions, and intelligence are believed to enable PLA forces to execute both concepts across the spectrum of conflict as well as to anticipate and mitigate any secondary effects.

PLA writings on war control emphasize the need to seize the initiative in the early stages of a conflict and maintain the initiative throughout the conflict.[19] Seizing the initiative at the outset of a conflict is believed to be crucial because it puts the opponent on the defensive, creating favorable conditions

AL: Air University, January 2022): 86–88, https://www.airuniversity.af.edu/CASI/Display/Article/2913216/in-their-own-words-2020-science-of-military-strategy/; Laird, *Managing Escalation*, 4; Morgan et al., *Dangerous Thresholds*, 51–54; and Alison Kaufman and Daniel Hartnett, *Managing Conflict: Examining Recent PLA Writings on Escalation Control* (Arlington, VA: CNA China Studies, February 2016), 11–12, https://www.cna.org/reports/2016/drm-2015-u-009963-final3.pdf.

13   China Aerospace Studies Institute, *PLA's Science of Military Strategy (2013)*, 13, 153; and Laird, *Managing Escalation*, 7, 9.

14   China Aerospace Studies Institute, *In Their Own Words*, 88; and Laird, *Managing Escalation*, 16–17.

15   China Aerospace Studies Institute, *PLA's Science of Military Strategy (2013)*, 112–13, 137–38; China Aerospace Studies Institute, *In Their Own Words*, 88; Laird, *Managing Escalation*, 5; Laird, *War Control*, 15, see Endnote 18; and Kaufman and Hartnett, *Managing Conflict*, 15.

16   Kaufman and Hartnett, *Managing Conflict*, 15.

17   Ibid., 15–16.

18   China Aerospace Studies Institute, *PLA's Science of Military Strategy (2013)*, 19–20; and China Aerospace Studies Institute, *In Their Own Words*, 51.

19   China Aerospace Studies Institute, *PLA's Science of Military Strategy (2013)*, 159–60; and China Aerospace Studies Institute, *In Their Own Words*, 91, 248, 263.

for limiting the conflict's scale, scope, pace, and intensity.[20] Once a conflict is underway, Chinese sources emphasize the need to continually seize the initiative at tactical and operational levels to maintain the advantage.[21]

Inexorable momentum calls for undertaking incrementally escalatory measures in a crisis or conflict to convince an opponent that, unless it backs down, China is certain to use decisive force.[22] Escalatory measures may include publicly stating a willingness to use force, raising alert levels, performing exercises, or launching limited attacks against strategic systems or informational nodes.[23] The concept of inexorable momentum supports war control by leveraging calculated, incremental escalation to precisely manipulate an opponent's perception of risk, making the opponent believe it cannot prevail and that the momentum of Chinese effort will make the opponent absorb increasingly undesirable costs. This momentum, in turn, helps control adversary behavior in a crisis or conflict by incentivizing concessions at the lowest possible level of force.

War control is ultimately rooted in the notion that war can and must be controlled across the spectrum of conflict, from the crisis phase through total war.[24] PLA writings on war control provide crucial insight into how Chinese strategists believe information technology, information-based capabilities, and intelligence can be leveraged to support the precise execution of warfare at all levels, including to control adversary perceptions and behavior.

ESCALATION RISKS ASSOCIATED WITH WAR CONTROL

PLA writings do not appear to address the potential escalation risks associated with war control, despite emphasizing the need to avoid unfavorable escalation.[25] U.S. planners must understand and account for these risks when formulating a tailored deterrence strategy vis-à-vis China.

Of particular concern are escalation risks associated with the PLA's emphasis on seizing the initiative, generating inexorable momentum, and leveraging advanced technologies to control war. Like China, U.S. operational doctrine also calls for seizing the initiative early in a conflict to gain the advantage. In a crisis or conflict between the two countries, escalatory steps taken by one or both sides to capitalize on a perceived window of opportunity may increase pressure on the other to respond in order to avoid being caught in an unfavorable position.[26] In other words, with both sides prioritizing swift, decisive action at the outset, the risk of miscalculation and of crossing key thresholds increases. This may, in turn, fuel inadvertent escalation.

Chinese leaders may also miscalculate an adversary's reaction to incremental escalation in the context of inexorable momentum. For instance, how can Chinese planners be certain that a limited attack against a space-based asset would successfully deter an adversary and contain, rather than catalyze, escalation? Would U.S. leaders interpret this as a limited signal or as something more

---

20   Morgan et al., *Dangerous Thresholds*, 51–53.

21   China Aerospace Studies Institute, *In Their Own Words*, 248.

22   Laird, *Managing Escalation*, 24; and Iain Alastair Johnston, "The Evolution of Interstate Security Crisis-Management Theory and Practice in China," *Naval War College Review 69*, no. 1 (Winter 2016): 20, https://core.ac.uk/download/pdf/236321806.pdf.

23   Laird, *Managing Escalation*, 24; and Johnston, "The Evolution of Interstate Security Crisis-Management Theory and Practice in China," 20.

24   China Aerospace Studies Institute, *In Their Own Words*, 140.

25   China Aerospace Studies Institute, *In Their Own Words*, 88–89, 118–119; and Laird, *War Control*, 6. Also see Kaufman and Hartnett, "Managing Conflict: Examining Recent PLA Writings on Escalation Control," 16.

26   Laird, *Managing Escalation*, 16.

escalatory? Indeed, even if Chinese strategists believe a specific action is limited and its effects can be perfectly controlled, there is always a risk that an adversary may interpret the action as highly threatening amid a crisis or conflict and decide to escalate.

A third escalation risk stems from the high-end technologies that are believed to enable war control. Systems that depend on AI and machine learning to predict with confidence what an opponent will do are only as accurate as their underlying algorithms and selected metrics. If the system's algorithms and metrics are skewed, leaders could draw fundamentally inaccurate conclusions about a situation that amplify their opportunism or desperation, culminating in decisions that lead to escalation. Take, for instance, the Soviet Union's VRYAN computer model that contributed to the Soviet "War Scare" in the early 1980s. The VRYAN program was designed to "objectively measure the correlation of forces and warn when Soviet relative strength had declined to the point that a preemptive Soviet attack might be justified."[27] According to a now-declassified intelligence assessment:

> KGB analysts working on VRYAN operated under the premise that the United States, when it had decisive overall superiority, might be inclined to launch an attack on the Soviet Union. In light of this assumption and because the program was supposed to determine, in a quantifiable way, when such a situation might be approaching, they believed it could provide strategic warning when the USSR was in a critically weak position relative to the United States, and conditions therefore were potentially conducive to a US attack.[28]

The assessment continues:

> By 1984 VRYAN calculated that Soviet power had actually declined to 45 percent of that of the United States. Forty percent was viewed as a critical threshold. Below this level, the Soviet Union would be considered dangerously inferior to the United States ... if the Soviet rating fell below 40 percent, the KGB and the military leadership would inform the political leadership that the security of the USSR could not be guaranteed ... the USSR would launch a preemptive attack within a few weeks of falling below the 40-percent mark.[29]

The VRYAN program was ultimately based on faulty assumptions about U.S. plans and decisionmaking. It is now clear that these underlying assumptions, when combined with the troves of data filtered through the computer model, fueled serious misperceptions that nearly caused Soviet leadership to escalate.[30] These technological risks are relevant to the PLA's contemporary war control concept. Although the 2013 SMS states that it is "necessary to prevent 'blind faith' in science and technology and 'the theory of the unique importance of weapons,' and [to avoid] exaggerating the impact that the key factor of science and technology has on military strategy and to regard this impact as absolute," PLA sources largely fail to acknowledge the escalation risks associated with the high-end technologies that are believed to enable the PLA to exercise strict control across the spectrum of conflict.[31] Failing to account for the possibility that modern technologies may perform imperfectly heightens the risk of miscalculation, which may drive escalation in a crisis or conflict.

---

27   President's Foreign Intelligence Advisory Board, *The Soviet 'War Scare'* (Washington, DC: February 1990): vi, https://nsarchive.gwu.edu/document/21038-4-pfiab-report-2012-0238-mr.

28   Ibid., 26.

29   Ibid., 45.

30   Ibid., 45–46.

31   China Aerospace Studies Institute, *PLA's Science of Military Strategy (2013)*, 19–20.

FRAMEWORK FOR STRENGTHENING DETERRENCE IN THE CONTEXT OF WAR CONTROL

As competition between the United States and China intensifies, the PLA's war control concept poses a unique yet pressing challenge for U.S. strategists. Indeed, the escalation risks associated with war control make strengthening deterrence during day-to-day competition imperative. This section outlines a framework for how the United States should think about and enhance deterrence in the context of war control.

To prevent a crisis or conflict with China, the United States should pursue measures during day-to-day competition that erode Chinese leaders' confidence in their ability to precisely plan for, execute, and control warfare across the spectrum of conflict. Recommendations to this end include:

1. STRENGTHEN DETERRENCE DURING DAY-TO-DAY COMPETITION.

An effective deterrence strategy that accounts for the PLA's war control concept must address *when* actions should be undertaken to achieve a strategic objective. The United States should take tailored and deliberate steps during the current period of competition to undermine Chinese leaders' confidence in their ability to exercise war control. PLA strategists emphasize acting only when the strategic situation is believed to be clear and favorable and when the process of military conflict can be controlled.[32] Increasing the PLA's perception of tactical, operational, and strategic uncertainty today may therefore make Chinese leaders less inclined to engage in provocative or opportunistic behavior that raises the risk of a crisis or conflict.

2. INFLUENCE PLA PERCEPTIONS OF CONTROL.

To undermine Chinese leaders' confidence in their ability to exercise war control, the United States should target key ways and means that underpin the PLA's war control concept. Here, key ways include the PLA's ability to seize the initiative and generate inexorable momentum. Key means include the capabilities and systems that enable the PLA to exercise seamless command and control, maintain information dominance, prepare strategic intelligence, and carry out precision strikes. Introducing uncertainty and unpredictability into these areas, in terms of PLA perceptions of their own capabilities as well as their perceptions of U.S. capabilities and responses, will bolster deterrence.

Broadly, U.S. actions should fall into two categories: actions taken to enhance deterrence by denial and actions taken to enhance deterrence by punishment. Deterrence by denial "seeks to deter an action by making it infeasible or unlikely to succeed," which in turn denies "a potential aggressor confidence in attaining its objectives."[33] Deterrence by punishment threatens to impose significant costs on an adversary if the adversary undertakes an unwanted action.[34] The United States can undermine PLA leaders' perception of their ability to exercise control by demonstrating an ability to deny and punish in new and unpredictable ways.

Another dimension of strengthening deterrence concerns the manner in which the United States signals or reveals its capabilities during day-to-day competition. In addition to overt signaling, such as deploying new platforms in theater or diversifying deployment patterns, the United States should consider strategically revealing covert capabilities to increase perceptions of risk and uncertainty in

---

32   China Aerospace Studies Institute, *In Their Own Words*, 64, 88.

33   Michael J. Mazaar, "Understanding Deterrence," RAND Corporation, 2018, 2, https://www.rand.org/pubs/perspectives/PE295.html.

34   Ibid., 2.

Beijing. Indeed, while the United States may possess certain capabilities that, through their capacity to deny or punish, could credibly undermine Chinese leaders' confidence in their ability to exercise war control, the covert nature of those capabilities limits their deterrent value. U.S. strategists, therefore, should evaluate which capabilities the United States may be able to reveal during day-to-day competition to inject uncertainty into the minds of PLA planners, making China doubt their assessment of U.S. forces and resolve as well as their ability to precisely control U.S. behavior across the spectrum of conflict.

Specifically, the United States could leverage overt signaling alongside the strategic revelation of covert capabilities to strengthen deterrence in the context of war control.

OVERT SIGNALING

"Overt signaling" refers to the strategic revelation of plans or capabilities that are not deliberately concealed. These measures can be both military and nonmilitary in nature. The aim of overt signaling should be to complicate the PLA's targeting and planning, specifically as it relates to the PLA's ability to seize the initiative and generate inexorable momentum through exercising seamless command and control, maintaining information dominance, leveraging strategic intelligence, and employing precision-strike assets. Overt military measures that will help strengthen deterrence can be seen in Table 1.

Table 1: Overt Military Signaling

Measure	Deterrence Type
Leverage open-source intelligence to detect and publicize PLA plans, movements, and capabilities.	Denial
Visibly distribute capabilities and expand key logistics nodes.	Denial
Overtly increase the redundancy and resiliency of key assets across domains.	Denial
Act in exercises in ways that are significantly different than how the PLA expects U.S. forces to act.	Punishment, Denial
Visibly deploy additional distributed strike assets in theater.	Punishment, Denial
Expand platform and theater missile defense capabilities.	Denial
Bolster security cooperation with allies and partners in the Indo-Pacific to enhance allied and partner capacity, readiness, and interoperability.	Punishment, Denial

Source: Author's research and analysis.

The United States should also consider how to leverage overt nonmilitary means to undermine PLA leaders' confidence in their ability to exercise war control. To this end, the United States should seek to increase PLA perceptions of uncertainty in economic and political domains through the measures in Table 2.

Table 2: Overt Nonmilitary Signaling

Measure	Deterrence Type
Demonstrate firm resolve and react to provocations in unpredictable ways. (For instance, the U.S. reaction to Russian aggression in Ukraine did not conform to Russian or Chinese expectations, which is important for undermining war control decision models.)	Punishment, Denial
Demonstrate an ability to threaten China's core economic objectives, in concert with allies and partners, in new and unpredictable ways.	Punishment, Denial
Bolster U.S. industrial capacity to enhance industrial resilience.	Denial
Enhance NATO resilience against Chinese political and economic subversion, including efforts to "control key technological and industrial sectors, critical infrastructure, and strategic materials and supply chains."[35]	Denial
Expand tailored diplomatic and development efforts to overtly counter Chinese information and influence operations abroad.	Denial

Source: Author's research and analysis.

COVERT CAPABILITY REVELATION

In 1989, RAND published a paper by Kevin N. Lewis introducing the idea of deliberate capability revelation (DCR). DCR refers to "the intentional release of authentic information about previously covert U.S. military capabilities, with the aim of manipulating adversary military balance assessments."[36] The basis of DCR is that "a covert capability has no deterrent value until some aspects of its nature are known."[37]

This section seeks to build on the existing body of literature that addresses DCR by outlining guiding points for covert capability revelation specifically in the context of Chinese war control.[38] While existing work on this topic focuses on the revelation of covert military capabilities, this paper considers the revelation of covert military *and* nonmilitary capabilities.

There are clear risks and limitations associated with covert capability revelation. First, revealing a previously covert capability will likely prompt adversary countermeasures. Effectively anticipating and managing the effects of adversary countermeasures so they do not negate a covert capability's advantages poses a complex challenge for U.S. strategists. Similarly, revealing parts of a capability may lead an adversary to uncover other design features, employment techniques, and tactics associated with that capability.[39] This may compromise the capability as a whole and enable the adversary to develop a similar asset.[40] The United States must therefore carefully guard other aspects of the capabilities it reveals to avoid lending the PLA any advantages. Finally, for covert capability

---

35   NATO, *NATO 2022 Strategic Concept (Brussels: June 2022)*, 5, https://www.nato.int/strategic-concept/.
36   Kevin N. Lewis, *Getting More Deterrence Out of Deliberate Capability Revelation* (Santa Monica, CA: RAND Corporation, August 1989): 3, https://www.rand.org/pubs/notes/N2873.html.
37   Ibid., v.
38   For more on this topic, see: Brendan Rittenhouse Green and Austin Long, "Conceal or Reveal? Managing Clandestine Military Capabilities in Peacetime Competition," *International Security* 44, no. 3 (Winter 2019/2020): 48–83, doi:10.1162/isec_a_00367; Thomas G. Mahnken, *Selective Disclosure: A Strategic Approach to Long-Term Competition* (Washington, DC: Center for Strategic and Budgetary Assessments, 2020): 1–35, https://csbaonline.org/research/publications/selective-disclosure-a-strategic-approach-to-long-term-competition; and Bernard Brodie, "Military Demonstration and Disclosure of New Weapons," *World Politics* 5, no. 3 (April 1953): 281–301, doi:10.2307/2009134.
39   Lewis, *Getting More Deterrence Out of Deliberate Capability Revelation*, 31.
40   Ibid.

revelation to be effective, the United States must understand Chinese leaders' knowledge f U.S. forces and nonmilitary capabilities as well as their "ignorance of our capabilities."[41] Inaccurate assessments of what the Chinese know, or do not know, may jeopardize covert capability revelation.

As with overt signaling, U.S. covert revelations during day-to-day competition should seek to influence Chinese leaders' perception of key war control ways (i.e., their ability to seize the initiative and generate inexorable momentum) and means (e.g., exercising seamless command and control, maintaining information dominance, gathering and leveraging strategic intelligence, and employing precision-strike assets). U.S. planners must also consider which measures would yield the most deterrent value if they could be directly attributed to the United States and which actions would yield the most deterrent value if they were nonattributable. Potential covert military capability revelations can be seen in Table 3.

Table 3: Covert Military Revelations

Measure	Deterrence Type
Reveal intelligence that undermines Chinese leaders' confidence in their ability to manipulate U.S. behavior and achieve surprise at strategic, operational, and tactical levels. This could include revealing intelligence that denies PLA leaders confidence in their ability to generate momentum or exploit perceived asymmetric advantages to seize the initiative.	Denial
Demonstrate an ability to employ or operate existing capabilities in novel ways, either to deny or punish, to complicate PLA planning.	Punishment, Denial
Demonstrate an ability to disrupt or defend against critical PLA systems in ways that were previously unknown to the PLA.	Denial
Reveal previously concealed redundancy and resiliency measures to increase PLA leaders' perception of uncertainty.	Denial
Reveal previously concealed operational concepts, campaign design, or tactics with the aim of creating additional operational uncertainty in the minds of PLA planners.	Denial

Source: Author's research and analysis.

Table 4 shows some potential covert nonmilitary capability revelations.

Table 4: Covert Nonmilitary Revelations

Measure	Deterrence Type
Demonstrate an ability to manipulate the nonmilitary information picture or counter Chinese information operations in a manner that undermines Chinese perceptions of information dominance.	Denial
Reveal novel and previously clandestine means of: (1) inflicting economic punishment against China or (2) defending against Chinese economic subversion to increase Chinese leaders' uncertainty regarding their ability to control economic objectives.	Punishment, Denial
Demonstrate an ability to interfere with Chinese diplomatic efforts and development projects abroad to increase political and diplomatic uncertainty.	Punishment, Denial

Source: Author's research and analysis.

---

41   Ibid., 6.

3. CONSIDER RISK-REDUCTION AND CRISIS-MANAGEMENT MEASURES IN THE CONTEXT OF WAR CONTROL.

As the United States aims to make war control appear more uncertain to Beijing, it should rigorously examine opportunities for minimizing miscalculation, misunderstanding, and miscommunication. However, the United States must be cognizant of Chinese perceptions of traditional risk-reduction and crisis-management measures. Indeed, China may seek to exploit risk-reduction and crisis-management mechanisms in a crisis or conflict to exercise control and gain a strategic advantage.[42]

Here, the United States' experience with U.S.-China hotline agreements is telling. The United States has signed two hotline agreements with China—one at the presidential level and one at the secretary-of-defense level.[43] On more than one occasion, U.S. officials have attempted to use established hotlines to contact their Chinese counterparts during a crisis, but Chinese leaders did not answer.[44] According to a 2022 RAND commentary:

"Hotlines are not meant to resolve the crisis but to empower higher-level organs within the PRC to signal resolve, assign blame, and stall until Beijing stakes out a position of maximum pressure and leverage over the United States during negotiations."[45]

Despite significant differences in the way the Chinese perceive and approach risk reduction and crisis management, examining how the United States may be able to reassure Chinese leadership in a crisis or conflict to mitigate escalation risks is essential.

KEY RISKS

This framework raises a number of risks worthy of consideration. First, pursuing any of the above measures during day-to-day competition may be perceived by Chinese leadership as escalatory. It is therefore imperative that U.S. strategists have a clear understanding of PLA thresholds and thinking on escalation. Second, some of the proposed measures may trigger adverse reactions from allies and partners. A core challenge will be determining whether and how to engage allies on some of the measures proposed above, with the aim of mitigating unfavorable reactions that may undermine alliance unity. Finally, some of the overt signaling measures outlined above may face opposition from the U.S. public and domestic interest groups. To prevent domestic opposition from undermining the perceived credibility of U.S. deterrent options, U.S. leaders should leverage strategic communication to articulate the threat picture and underscore why taking steps today to enhance deterrence vis-à-vis China is imperative.

CONCLUSION

The Chinese war control concept raises complex challenges for U.S. deterrence planners. The United States must acknowledge the risks associated with war control and tailor its deterrence strategy accordingly. Some may dismiss the war control concept by asserting that the PLA does not have the capability or capacity to strictly control activities across the spectrum of conflict. However,

---

42   Lyle J. Morris and Kyle Marcrum, "Another 'Hotline' with China Isn't the Answer," RAND Corporation, July 27, 2022, https://www.rand.org/blog/2022/07/another-hotline-with-china-isnt-the-answer.html.

43   Ibid.

44   Ibid; and Phelim Kine, "'Spiral into crisis': The U.S.-China military hotline is dangerously broken," *Politico*, September 1, 2021, https://www.politico.com/news/2021/09/01/us-china-military-hotline-508140.

45   Morris and Marcrum, "Another 'Hotline' with China Isn't the Answer."

irrespective of whether the PLA has the capability to exercise war control, the crucial factor is whether Chinese leaders *believe* they can exercise control across the spectrum of conflict. To help deter a major crisis or armed conflict with China, the United States must take steps today to convince Chinese leadership that it cannot obtain the core objectives of war control.

Kim Jong-Un's Nuclear To-Do List

Contextualizing North Korean Missile Tests, 2021–Present

By Jaejoon Kim[1]

INTRODUCTION

In November 2017, Kim Jong-un oversaw North Korea's second intercontinental ballistic missile (ICBM) test and subsequently declared his nuclear deterrent against the United States to be complete.[2] After a notable absence of testing in 2018, North Korea resumed missile tests by firing multiple short-range ballistic missiles (SRBMs) in May 2019. Pyongyang conducted 15 missile tests between May 2019 and July 2020 of various types and ranges, followed by another lull in testing. However, it has again resumed testing, having overseen 42 tests since January 2021.

North Korean missile tests have once again become a regular occurrence. The hopes for long-term détente between the United States and North Korea have all but dissipated, while calls for additional dialogue from Washington and Seoul only return hostility from Pyongyang. Despite the "completion" of North Korea's nuclear deterrent in late 2017, following the successful tests of the Hwasong-14 and Hwasong-15 ICBMs, Kim Jong-un has also acknowledged that there is no shortage of improvements to the nuclear arsenal that must be made to maximize nuclear leverage against the United States and South Korea.[3] Indeed, if Kim was genuinely confident in the deterrent, there would be no need for further missile tests. This paper examines how recent missile tests contribute to Kim Jong-un's

1   Jaejoon Kim is an intelligence specialist at Strider Technologies, Inc. The views expressed in this paper are those of the author and do not reflect the views of Strider or its staff, board, or clients.

2   "Kim Jong Un's 2018 New Year's Address," National Committee on North Korea, January 1, 2018, https://www.ncnk.org/node/1427.

3   "FM Spokesman Flails U.S. Anti-DPRK Moves over Cyber Issue," Korean Central News Agency (KCNA), December 21, 2017, https://kcnawatch.org/newstream/1513852222-173114421/fm-spokesman-flails-u-s-anti-dprk-moves-over-cyber-issue/.

broader goals and further details the improvements he seeks. In other words, this paper enumerates Kim's "to-do list" when it comes to North Korea's nuclear forces.

METHODOLOGY

The closest primary source that exists for Pyongyang's nuclear aspirations comes from a January 2021 report given by Kim Jong-un himself to the Worker's Party of Korea (WPK) at the 8th Party Congress. Titled "Great Programme for Struggle Leading Korean-Style Socialist Construction to Fresh Victory On Report Made by Supreme Leader Kim Jong Un at Eighth Congress of WPK" (henceforth "The Great Program"), the report was a de facto substitute for Kim's annual New Year's speech. However, while the New Year's speeches tend to reiterate basic talking points and vague platitudes on the glory of the Juche ideology, the Great Program addresses a wide variety of policy sectors in a highly granular fashion. Of particular importance for this study are the five capabilities related to North Korea's nuclear arsenal that Kim outlined for further progress:

1. hypersonic glide vehicles (HGVs);

2. multiple reentry vehicles (MRVs);

3. solid-fuel ICBMs;

4. tactical nuclear weapons (TNWs); and

5. undersea strategic forces, including nuclear-powered ballistic missile submarines (SSBNs) and submarine-launched ballistic missiles (SLBMs).

Progress has varied across the capabilities since the Great Program's publication, implying that some are more feasible or urgent to the North Korean nuclear agenda than others.

Given Kim's centrality to the agendas of both the WPK and the Korean People's Army (KPA), this analysis considers his public statements to be an authoritative source for the direction of North Korea's arsenal. The Great Program's recency and attention to detail make it a strong reference point for the arsenal's future direction. This piece will: (1) identify the five capabilities in the Great Program, (2) connect North Korea's missile tests since January 2021 to those five capabilities, and (3) analyze North Korean progress on each of those items.

This paper treats the recent wave of missile tests as not reactionary (i.e., anchored to external developments that Pyongyang perceives as detrimental to its security interests) but rather as a means to measure progress toward reaching Kim's goals as outlined in the Great Program. This is not to dismiss the possibility of missile launches as a direct response to joint military exercises between the United States and South Korea, which absolutely occur.[4] Rather, it acknowledges the possibility that missile tests can always serve more value than merely a response to unfavorable circumstances.

Finally, it is worth noting that the ongoing opacity surrounding North Korean nuclear goals and decisionmaking is a source of strength for Pyongyang, as it increases the likelihood that analysts draw varying conclusions about Kim's intent—even as they are drawn from the same primary sources.[5]

---

4   "Report of General Staff of KPA on Its Military Operations Corresponding to U.S.-South Korea Combined Air Drill," Korean Central News Agency, November 7, 2022, https://kcnawatch.org/newstream/1667774164-903270856/report-of-general-staff-of-kpa-on-its-military-operations-corresponding-to-u-s-south-korea-combined-air-drill.

5   Toby Dalton, Narushige Michishita, and Tong Zhao, "Security Spillover: Regional Implications of Evolving Deterrence on the Korean Peninsula," Carnegie Endowment for International Peace, June 11, 2018, https://carnegieendowment.org/2018/06/11/security-spillover-regional-implications-of-evolving-deterrence-on-korean-peninsula-pub-76483.

A threat that can be understood is a threat than can be countered, and Pyongyang is likely to continue to use its unpredictability to impede its enemies from adopting the correct response; the conclusions drawn from this piece are not immune to this reality.

THE FIVE CAPABILITIES

The Great Program is reported to have taken Kim Jong-un a staggering nine hours to read. Excerpts relevant for this analysis are presented below, bolded for emphasis:

> The report also noted that in the period under review the sector of national defence scientific research was conducting research into perfecting the guidance technology for **multi-warhead rocket** at the final stage, finished research into developing warheads of different combat missions including the **hypersonic gliding flight warheads** for new-type ballistic rockets and was making preparations for their test manufacture.
>
> …
>
> And the tasks were brought up to develop and introduce **hypersonic gliding flight warheads** in a short period, push ahead with the development of **solid-fuel engine-propelled intercontinental underwater and ground ballistic rockets** as scheduled, and possess a **nuclear-powered submarine and an underwater-launch nuclear strategic weapon** which will be of great importance in raising the long-range nuclear striking capability.[6]

The two paragraphs above may not comprehensively capture Kim's nuclear ambitions, but they nevertheless provide useful insight on the direction that the arsenal is expected to take in the future.

1. HYPERSONIC GLIDE VEHICLES (HGVS)

While most ballistic missiles travel at hypersonic speeds—five times the speed of sound—HGVs' distinguishing feature is their ability to travel and maneuver on a non-ballistic flight trajectory (a ballistic trajectory refers to a parabolic flight path in which a missile depletes its fuel during ascent and relies solely on gravity for its descent).[7]

Often considered a capability reserved for highly advanced and well-established nuclear powers, HGVs were not thought to be a feasible project for North Korea when the Great Program was first made public. Even as reports confirmed the establishment of a research center for hypersonic missiles capable of bypassing missile defense systems, North Korea's pursuit of HGVs was written off as wishful thinking or an exaggerated threat designed for the Great Program's foreign audience.[8]

Yet contrary to expectations, North Korea confirmed that it had tested the Hwasong-8 "hypersonic missile" in September 2021, just nine months after the reports of initial research and development.[9]

---

6   "Great Programme for Struggle Leading Korean-Style Socialist Construction to Fresh Victory On Report Made by Supreme Leader Kim Jong Un at Eighth Congress of WPK," KCNA, January 10, 2021, https://kcnawatch.org/new-stream/1610261416-871234007/great-programme-for-struggle-leading-korean-style-socialist-construction-to-fresh-victory-on-report-made-by-supreme-leader-kim-jong-un-at-eighth-congress-of-wpk/.

7   Kelley M. Sayler and Amy F. Woolf, "Defense Primer: Hypersonic Boost-Glide Weapons," Congressional Research Service, November 17, 2021, https://sgp.fas.org/crs/natsec/IF11459.pdf.

8   Tae Joo Jeong, "North Korea Forms New Research Center Focused on 'Hypersonic Missiles,'" Daily NK, January 6, 2021, https://www.dailynk.com/english/north-korea-forms-new-research-center-focused-on-hypersonic-missiles/; and Won-Gi Jung, "North Korea's Promised Military Upgrades Are a Tall Order — and an Empty Threat," NK News, January 9, 2021, https://www.nknews.org/2021/01/north-koreas-promised-military-upgrades-are-a-tall-order-and-an-empty-threat/?t=1634055683699.

9   "Academy of Defence Science test-fires a new type of hypersonic missile," KCNA Watch, September 29, 2021, https://kc-

After two additional tests in 2022, it is still unclear if North Korea possesses a true boost-glide vehicle, a simpler maneuverable reentry vehicle (MaRV), or both.[10] Equally unclear is the missile's usability in battle, which North Korean state media has an incentive to inflate. At any rate, the astonishing pace with which Pyongyang has made good on its promise—at least at the surface level—indicates that the five capabilities outlined in the Great Program are far cries from mere propaganda and that Kim may be serious about achieving mastery over all of them. However, the Hwasong-8 has yet to be tested since January 2022, despite its impressive debut; it remains to be seen whether more tests are forthcoming or if progress on this particular item has halted.

Given the similarity of the Hwasong-8's engine to that of the Hwasong-12 intermediate-range ballistic missile (IRBM)—a North Korean missile capable of striking Guam—the Hwasong-8 could potentially also target U.S. installations there; that said, differing payload sizes between the Hwasong-8 and Hwasong-12 make it difficult to assess the former's intended range with much confidence.[11]

What is clearer is that the Hwasong-8 was likely developed with the intent to overwhelm U.S. theater missile defense installations in East Asia, such as Terminal High-Altitude Area Defense (THAAD) installations.[12] HGVs' non-ballistic flight trajectories allow them to evade detection and interception from systems such as THAAD during the midcourse phase. An important caveat is that the protection that HGVs offer during the midcourse phase may incur a trade-off with protection during the terminal phase.[13]

2. MULTIPLE REENTRY VEHICLES (MRVS)

If HGVs are North Korea's answer to theater missile defense, Pyongyang has likely found its response to U.S. homeland missile defense in MRVs. The United States currently deploys the Ground-based Midcourse Defense (GMD) system to repel a surprise strike from North Korea. The logic of the GMD is that a sufficiently small strike on the United States—one befitting of a nuclear rogue state such as North Korea but not a nuclear power such as China or Russia—can be totally rebuffed by GMD interceptors based in Alaska and California before any missiles strike the U.S. homeland.[14]

This logic, however, relies on the assumption that each missile in each strike can only carry one warhead. If incoming North Korean missiles were to have more than one reentry vehicle (RV) attached to them, they would spread U.S. interceptors thin and drastically increase the odds that one of those RVs successfully detonates, imposing unacceptable costs on the United States. The ability of MRVs to strain U.S. homeland missile defense—and by extension, augment North Korea's nuclear deterrent—makes it highly desirable for Pyongyang.

Aside from numerous and widely discussed structural problems with GMD, Pyongyang has been hard at work to overcome Washington's missile shield.[15] On October 2020, North Korea debuted a

nawatch.org/newstream/1632906065-289775574/academy-of-defence-science-test-fires-a-new-type-of-hypersonic-mis-sile/?t=1640706089909.

10   Vann H. Van Diepen, "Another North Korean 'Hypersonic' Missile?," 38 North, January 7, 2022, https://www.38north. org/2022/01/another-north-korean-hypersonic-missile/.

11   Ibid.

12   Jeong, "North Korea Forms New Research Center Focused on 'Hypersonic Missiles.'"

13   Cameron L. Tracy and David Wright, "Modeling the Performance of Hypersonic Boost-Glide Vehicles," *Science & Global Security* 28, no. 3 (2020): 135–170, doi:10.1080/08929882.2020.1864945.

14   Leo Bryne, "U.S. conducts "critical" missile defense test of GMD system," NK News, March 25, 2019, https://www.nknews. org/2019/03/u-s-conducts-critical-missile-defense-test-of-gmd-system/.

15   Jaganath Sankaran and Steve Fetter, "Defending the United States: Revisiting National Missile Defense against North Korea," *International Security* 46, no. 3 (Winter 2021/2022): 51–86, doi:10.1162/isec_a_00426; and Ankit Panda, "North

missile even larger than the Hwasong-15 ICBM, a missile which is already capable of ranging the entire continental United States. This led experts to conclude that the new missile, later dubbed the Hwasong-17, is designed to be an ICBM with a higher payload, which strongly implies that the missile is MRV-capable.[16]

Additionally, North Korea's recent SLBM models—the Pukguksong-3, Pukguksong-5, and an even larger unnamed SLBM unveiled at a military parade on April 2022—are believed to be capable of lofting multiple MRVs as well.[17] In this context, the possession of MRVs provides additional insurance against U.S. homeland missile defense. That said, it is entirely possible that North Korea could equip a theater missile with multiple warheads at the expense of range.

North Korea has yet to conduct a test with an MRV-equipped missile, perhaps because North Korea has yet to perfect even single-warhead reentry on its ICBMs.[18] Once Pyongyang is comfortable with reentry, however, there are strong incentives for future missile tests to be conducted with MRVs. For starters, mastery over MRVs would push Pyongyang closer to multiple *independent* reentry vehicles (MIRVs) capable of targeting multiple points as opposed to indiscriminately dropping submunitions, further raising the difficulty of countering North Korean missiles using the GMD. Additionally, independent of the augmented threat value that MRVs or MIRVs have against Seoul and Washington, the ability to test multiple warheads with a single missile test would multiply precious testing results and findings with which Pyongyang can further refine its arsenal.[19] None of these developments are favorable for the United States or South Korea.

3. SOLID-FUEL ICBMS

The current North Korean ICBM arsenal is exclusively liquid-fueled, even as its theater missiles have been updated with solid-fuel entries.[20] While liquid fuel generates more thrust per propellant mass (and thus provides longer range per unit), solid-fuel missiles are ready to launch upon manufacture and do not require on-site fueling, obviating the need for support vehicles and thereby making platforms less likely to be detected by early warning systems.[21] The relative ease of launch and small footprint of solid-fuel missiles vis-à-vis their liquid counterparts make the former far more attractive to Kim Jong-un.

North Korea has yet to test a solid-fuel ICBM, and the only SLBM that it tested in the past two years is understood to be a sea-launched variant of the KN-23 SRBM, implying that progress on a long-range solid-fuel missile has been slow.[22] Furthermore, Pyongyang has stated in recent months that it sees

Korea and U.S. Missile Defense Capabilities," National Bureau of Asian Research, September 8, 2021, https://www.nbr.org/wp-content/uploads/pdfs/publications/north_korea_brief_090821.pdf.

16   Christy Lee, "Experts: North Korea's New ICBM May Carry Multiple Nuclear Warheads," VOA News, March 25, 2022, https://www.voanews.com/a/experts-north-korea-s-new-icbm-may-carry-multiple-nuclear-warheads-/6501026.html.

17   Markus V. Garlauskas, "We Must Prevent North Korea from Testing Multiple Reentry Vehicles," *Beyond Parallel*, CSIS, November 5, 2020, https://beyondparallel.csis.org/we-must-prevent-north-korea-from-testing-multiple-re-entry-vehicles/.

18   Ankit Panda, Kim Jong Un and the Bomb: Survival and Deterrence in North Korea (Oxford, UK: Hurst Publishers, 2020): 217–222.

19   Garlauskas, "We Must Prevent North Korea from Testing Multiple Reentry Vehicles."

20   For more information, see https://missilethreat.csis.org/country/dprk/.

21   Ankit Panda, "North Korea Tests Its Pukkuksong-2 Solid Fuel Ballistic Missile Again: First Takeaways," *The Diplomat*, May 22, 2017, https://thediplomat.com/2017/05/north-korea-tests-its-pukkuksong-2-solid-fuel-ballistic-missile-again-first-takeaways/.

22   Vann H. Van Diepen, "North Korea's 'New Type Submarine-Launched Ballistic Missile': More Political Than Military Significance," 38 North, October 22, 2021, https://www.38north.org/2021/10/north-koreas-new-type-submarine-launched-ballistic-missile-more-political-than-military-significance/.

merit in "turning all missile systems into ampoules," or "ampoulizing" them.[23] Ampoulization essentially seals the liquid propellant and oxidizer tanks within the missile, enabling them to be pre-fueled at the factory rather than on the launchpad.[24] Assuming that Pyongyang possesses the capability to ampoulize its missiles, its statement suggests that it is in no rush to transition into an all-solid arsenal.[25] The recent round of Hwasong-17 tests in April further corroborates this hypothesis.

4. TACTICAL NUCLEAR WEAPONS (TNWS)

Currently, there is no universally accepted definition for a TNW. Indeed, it is nearly impossible to define precisely when a nuclear weapon becomes weak enough to only turn the tide of a single battle as opposed to a broader military campaign. "Yield" merely refers to a unit of energy; even a 1-kiloton nuclear warhead—among the smallest of its kind—could have tremendous implications on the entire planet, let alone the tide of a war, depending on where and when it is detonated. The blurred lines between "tactical" and "strategic" nuclear weapons make deciphering North Korean intent on its own "tactical nuclear weapons" even more challenging.

TNWs do not receive a mention in the two above excerpts from the Great Program but can instead be found in the passages below (bolded for emphasis):

> The national defence science sector developed the super-large MLRS, a super-power attack weapon the world's weaponry field had never known, and proceeded to develop **ultra-modern tactical nuclear weapons** including new-type tactical rockets and intermediate-range cruise missiles whose conventional warheads are the most powerful in the world.
>
> . . .
>
> It is necessary to develop the nuclear technology to a higher level and make nuclear weapons smaller and lighter for more tactical uses. This will make it possible to develop **tactical nuclear weapons** to be used as various means according to the purposes of operational duty and targets of strike in modern warfare, and continuously push ahead with the production of super-sized nuclear warheads. In this way we will be able to thoroughly contain, control and handle on our own initiative various military threats on the Korean peninsula, which are inevitably accompanied the nuclear threat.[26]

Additionally, the first paragraph deserves further attention due to a potential translation discrepancy discovered by the author during preliminary research. The original Korean text of the same report, published a day earlier by North Korean state media, can be seen below alongside the author's translations (punctuations and emphasis added for clarity):

23   "Hypersonic Missile Newly Developed by Academy of Defence Science Test-Fired," KCNA, September 29, 2021, https://kcnawatch.org/newstream/1632870069-293810660/hypersonic-missile-newly-developed-by-academy-of-defence-science-test-fired/.

24   Josh Smith, "Analysis: N.Korea looks to risky pre-fuelled missiles to reduce launch time," Reuters, January 11, 2022, https://www.reuters.com/world/asia-pacific/nkorea-looks-risky-pre-fuelled-missiles-reduce-launch-time-2022-01-11/.

25   Vann H. Van Diepen, "Six Takeaways From North Korea's 'Hypersonic Missile' Announcement," 38 North, October 13, 2021, https://www.38north.org/2021/10/six-takeaways-from-north-koreas-hypersonic-missile-announcement/.

26   "Great Programme for Struggle Leading Korean-Style Socialist Construction to Fresh Victory On Report Made by Supreme Leader Kim Jong Un at Eighth Congress of WPK," KCNA.

Official English Translation	Original Korean Text	Author Translation
The national defence science sector developed the super-large MLRS, a super-power attack weapon the world's weaponry field had never known, and proceeded to develop ultra-modern tactical nuclear weapons including new-type tactical rockets and intermediate-range cruise missiles whose conventional warheads are the most powerful in the world. This enabled us to gain a reliable edge in military technology.	국방과학부문에서 세계병기분야에서 개념조차 없던 초강력다련발공격무기인 초대형방사포를 개발완성하고 상용탄두위력이 세계를 압도하는 신형전술로케트와 중장거리순항미싸일을 비롯한 첨단핵전술무기들도 련이어 개발함으로써 믿음직한 군사기술적강세를 틀어줘였다.[27]	The national defense science sector developed the super-large MLRS, a weapon that has yet to be even conceived of by the other militaries of the world. Additionally, by developing new-type tactical missiles—whose conventional warheads are the most powerful in the world—and ultra-modern tactical nuclear weapons such as the intermediate-range cruise missile in rapid succession, the national defense science sector has granted us a reliable military advantage.

The author stresses that the new translation is not decisive. That said, the newly translated passage departs from the original translation in that it makes a clear distinction between "conventional new-type tactical missiles" and "nuclear intermediate-range cruise missiles," meaning that the intermediate-range cruise missile fired on September 2021 is meant to carry a nuclear payload.

As for the conventional "new-type" tactical missiles, Kim may be referring to either the KN-23, the most recent test of which measured its conventional firepower (but is likely dual-capable and not "new-type" as of 2021), or the close-range ballistic missile that was tested on April 16, 2022 (which may also be dual-capable but is certainly "new-type").[28] The lack of compelling evidence separating "conventional new-type tactical missiles" and "ultra-modern tactical nuclear weapons" in North Korea's known arsenal presents reason to give pause to the new translation.

5. STRATEGIC SUBMARINE FORCES (SSBNS AND SLBMS)

States with a nuclear submarine force have two basic options for how to use them: a bastion model that keeps the submarines close to port and deploys them ad hoc (as the Soviet Union did) or a continuous at-sea model which keeps the vessels on constant patrol (as the United States does).[29]

27   "우리 식 사회주의건설을 새 승리에로 인도하는 위대한 투쟁강령 조선로동당 제8차대회에서 하신 경애하는 김정은 동지의 보고에 대하여" [Great Programme for Struggle Leading Korean-Style Socialist Construction to Fresh Victory On Report Made by Supreme Leader Kim Jong Un at Eighth Congress of WPK], Korean Central News Agency, January 9, 2021, https://kcnawatch.org/newstream/1610179272-556433452/%EC%9A%B0%EB%A6%AC-%EC%8B%9D-%EC%82%AC%ED%9A%8C%EC%A3%BC%EC%9D%98%EA%B1%B4%EC%84%A4%EC%9D%84-%EC%83%88-%EC%8A%B9%EB%A6%AC%EC%97%90%EB%A1%9C-%EC%9D%B8%EB%8F%84%ED%95%98%EB%8A%94-%EC%9C%84%EB%8C%80/.

28   "Academy of Defence Science Conducts Important Weapons Tests," *Rodong Sinmun*, January 28, 2022, https://kcnawatch.org/newstream/1643345669-239469558/academy-of-defence-science-conducts-important-weapons-tests/; and William Gallo, "North Korea Fires Nuclear-Capable 'Tactical Guided Weapon'," VOA News, April 16, 2022, https://www.voanews.com/a/north-korea-fires-nuclear-capable-tactical-guided-weapon-/6532813.html.

29   Vipin Narang and Ankit Panda, "Command and Control in North Korea: What a Nuclear Launch Might Look Like," War on

A strategic submarine force, in theory, would provide North Korea with yet another second-strike capability and a highly effective shield against adversaries' nuclear coercion. That said, North Korea's progress on SSBNs and SLBMs has been the slowest out of all five of the capabilities outlined in the Great Program. While other missile tests have spiked in frequency, the testing rate for SLBMs has remained stable at best. The North Korean submarines capable of firing ballistic missiles are diesel electric SSBs, not nuclear-powered SSBNs. The Korea Chair at CSIS regularly publishes satellite imagery reports on the Sinpo and Nampo Shipyards whenever they detect unusual activity; the images, as of now, do not present a clear picture of progress on SSBNs and SLBMs.

Even if North Korea was able to eventually deploy a true SSBN fleet, the practical value that it would add to Pyongyang's nuclear deterrent is dubious. Developing SSBNs quiet enough to avoid acoustic detection in the Pacific Ocean would be a monumental task for Pyongyang. Moreover, a credible and survivable sea-launched second-strike capability will require adequate nuclear command, control, and communications (NC3) to properly receive and execute strike orders during wartime and—more importantly—ensure that the missiles are *not* fired during peacetime.[30] While not much is known about the extent of North Korean NC3, there is reason to believe that at-sea NC3 is and will remain a particularly difficult challenge for Pyongyang to overcome.[31]

In conclusion, strategic submarine forces remain an impractical means for North Korea to bolster its nuclear deterrent; the relative lack of observable activity in this front implies that these forces are not an urgent priority for Pyongyang.

NORTH KOREAN NUCLEAR GOALS

Given North Korea's notorious opacity and unpredictability, it is rarely prudent to claim knowledge of the state's strategic aspirations. However, it is possible to paint with a very broad brush and identify three non-mutually exclusive purposes behind Pyongyang's nuclear forces: regime survival, U.S.-ROK alliance decoupling, and battlefield nuclear use for establishing escalation dominance.

1. REGIME SURVIVAL: DETERRENCE BY PUNISHMENT

In April 2013, the Supreme People's Assembly of the WPK adopted the "Law on Consolidating Position of Nuclear Weapons State for Self-Defence," providing foreign audiences with a de facto (but decidedly not de jure) nuclear doctrine.[32] It declared the supreme commander of the KPA (Kim Jong-un) to possess sole authority over nuclear launch and that nuclear weapons would only be used under the following conditions:

- North Korea is attacked or invaded by a hostile nuclear weapons state; or

- North Korea is attacked or invaded by a non-nuclear state that has joined forces with a hostile nuclear state.

<hr>

the Rocks, September 15, 2017, https://warontherocks.com/2017/09/command-and-control-in-north-korea-what-a-nuclear-launch-might-look-like/.

30   Lami Kim, "A Race for Nuclear-Powered Submarines on the Korean Peninsula?," Maritime Awareness Project, National Bureau of Asian Research, March 31, 2021, https://www.nbr.org/publication/a-race-for-nuclear-powered-submarines-on-the-korean-peninsula/.

31   Panda, Kim Jong Un and the Bomb, 236–239.

32   "Law on Consolidating Position of Nuclear Weapons State Adopted," KCNA, April 1, 2013, https://kcnawatch.org/newstream/1451896124-739013370/law-on-consolidating-position-of-/Leitenberg/.

In September 2022, however, North Korean state media reported that the Supreme People's Assembly invalidated the 2013 law in favor of the "Law on Policy of Nuclear Forces."[33] The new law provides a longer list of circumstances that justify nuclear use, including:

- when an attack by nuclear weapons (or other weapons of mass destruction) has been launched at the homeland or is judged to be imminent;

- when an attack on North Korean leadership or the command structure of its nuclear forces is launched or judged to be imminent;

- when an attack on key strategic assets is launched or judged to be imminent;

- when a conflict has already erupted and North Korea must take the initiative to avoid the conflict's protraction or expansion; and

- in all other catastrophic circumstances when nuclear weapons use alone would provide safety to the North Korean state, people, and leadership.

Additionally, the new law inserts a clause on launch authority: if nuclear command and control is in jeopardy due to hostile attacks (e.g., a decapitation strike on Kim Jong-un), a nuclear strike will be *immediately and automatically* launched against the responsible party.[34]

Aside from the changes to launch authority, the new law on nuclear use does not present fundamental changes to the 2013 iteration; North Korea's informal nuclear doctrine remains ostensibly defensive. North Korea has been consistent and transparent in its messaging to the international community that its nuclear arsenal is intimately tied to the protection of the nation from potential invasion or other attacks on its sovereignty. During the 2018 Singapore Summit, Pyongyang cited the case of Libya under Muammar al-Qaddafi multiple times as a counter to Washington's calls for denuclearization.[35] The connection between Qaddafi's decision to give up its nascent nuclear arsenal in 2003 and his deposal in 2011 has not been—and will likely never be—lost on Kim Jong-un, who took power just months after Qaddafi's removal. Nuclear weapons are the *sine qua non* to North Korea's defense against outsiders and, by extension, the Kim family's grip on power. Crucially, even if the nuclear program is not necessary for North Korean survival and prosperity, Kim Jong-un certainly believes it to be and will continue to behave accordingly.

The nuclear arsenal not only defends against external forces, but also serves to bolster the regime's credibility against internal political threats. While Kim Jong-un's reign has been stable in recent years, there is precedent for North Korean nuclear advancements being used as a tool to secure the support of domestic elite factions within Pyongyang for the Kim dynasty.[36] North Korea's nuclear success story presents Kim as a powerful leader, and future strains to his credibility may be met with more dynamic movements in the nuclear field.[37]

---

33 "DPRK's Law on Policy of Nuclear Forces Promulgated," KCNA, September 9, 2022, https://kcnawatch.org/new-stream/1662721725-307939464/dprk%E2%80%99s-law-on-policy-of-nuclear-forces-promulgated/.

34 Ibid.

35 Megan Specia and David E. Sanger, "How the 'Libya Model' Became a Sticking Point in North Korea Nuclear Talks," *New York Times*, May 16, 2018, https://www.nytimes.com/2018/05/16/world/asia/north-korea-libya-model.html.

36 Edward Howell, "The Juche H-Bomb? North Korea, Nuclear Weapons and Regime-State Survival," *International Affairs* 96, no. 4 (2020): 1063–1064, doi:10.1093/ia/iiz253.

37 Bruce Bennett et al., *Countering the Risks of North Korean Nuclear Weapons* (Santa Monica, CA: RAND Corporation, 2021): 8–9, https://www.rand.org/pubs/perspectives/PEA1015-1.html.

2. U.S.-ROK ALLIANCE DECOUPLING

North Korea views South Korea as an extension of the security threat from the United States, not as a threat on its own. This patronizing narrative advances Pyongyang's interests by (1) reducing Seoul's regional role as a junior partner to Washington, (2) enabling Pyongyang to purport peaceful intent for intra-peninsular affairs, and (3) justifying the buildup and tests of short-range arms that would be used against South Korea.

Scholars have noted that North Korean advancements in its nuclear weapons program force South Korea to doubt the credibility of the U.S. nuclear umbrella, a foundational aspect of the security alliance between the two countries.[38] By continuing advancements in its intercontinental strike capabilities, Pyongyang can sow skepticism in both Washington and Seoul as to whether the United States is willing to risk a nuclear attack on its own cities to defend a country on the other side of the world.

This paper assumes that any advancement in North Korean nuclear capabilities constitutes a wedge between the U.S.-ROK alliance; improvements in long-range strategic capabilities such as ICBMs raise the stakes on U.S. regional involvement, while improvements in short-range capabilities drive South Korea to perceive North Korean aggression as imminent, forcing Seoul to call into question the stability of the security guarantee.

3. BATTLEFIELD USE AND ESCALATION DOMINANCE

Only a small portion of North Korea's nuclear arsenal is "strategic" in the traditional sense. Indeed, if North Korea could only respond to crises in the Korean Peninsula with an all-out nuclear war against the United States and South Korea, it too would inevitably be annihilated by the U.S. response. The "threat that leaves something to chance," to borrow Thomas Schelling's words, is ultimately one that is not credible; the North Korean response would essentially reduce to selecting between surrender or suicide.[39]

Recent developments show that North Korea is deeply interested in equipping itself with a nuclear arsenal that can realistically be used to *win* wars, not just deter adversaries. Past statements from Pyongyang show that many of its intended targets for its nuclear weapons are military installations in South Korea, Japan, and U.S.-controlled Guam.[40] The missiles that North Korea has tested in the past year would be able to strike all of them.

During a large-scale military parade in April 2022, Kim stated: "If any forces try to violate the fundamental interests of our state, our nuclear forces will have to decisively accomplish its unexpected second mission."[41] Those who study North Korea are no strangers to cryptic messaging from Kim Jong-un; however, the timing of such a message, coinciding with one of the fiercest chapters of the Russian invasion of Ukraine, indicates that Kim has been watching Vladimir Putin's deployment of nuclear coercion closely. Putin's nuclear threats have been at least partially (if not wholly) successful in preventing direct intervention from the United States and its NATO allies in

---

38   Bennett et al., 50–51.

39   Robert Powell, "The Theoretical Foundations of Nuclear Deterrence," *Political Science Quarterly* 100, no. 1 (Spring 1985): 78, doi:10.2307/2150861.

40   Matt Korda, "Nuclear Weapons and Delivery Systems That Might Be Implicated in Nuclear Use Involving the Korean Peninsula," *Journal for Peace and Nuclear Disarmament* 5, supp 1 (April 2022), doi:10.1080/25751654.2022.2055911.

41   "Respected Comrade Kim Jong Un Makes Speech at Military Parade Held in Celebration of 90th Founding Anniversary of KPRA," KCNA, April 26, 2022, https://kcnawatch.org/newstream/1650963237-449932111/respected-comrade-kim-jong-un-makes-speech-at-military-parade-held-in-celebration-of-90th-founding-anniversary-of-kpra/.

Ukraine.[42] Kim may believe that mastery over battlefield nuclear use would convince the United States and Japan to stay out of the Korean Peninsula. At the rate that he has been testing his missiles, mastery may not be so far away.

MAPPING THE TESTS

Historical accounts differ in the precise number of tests that North Korea has conducted since the release of the Great Program in January 2021, with some counting tests as military exercises and vice versa. This paper counts 42 missile tests by North Korea since January 2021 at time of writing. The tests in recent months cannot be reduced exclusively to their contributions to the five military capabilities; missile tests serve a litany of important functions for Pyongyang beyond assessing the viability of isolated military systems. Alongside the messages they send to its adversaries, missile tests also provide the KPA with valuable information for further refining its capabilities.

Information gleaned from a single type of missile spills over into other missile types. For example, the lessons learned from the testing of the Hwasong-8 will almost certainly be applicable to future improvements to the Hwasong-12, given the similarity of their boosters.[43] Below, this paper compiles such testing data to evaluate their contributions to the five capabilities in the Great Program.[44] This analysis includes the following assumptions:

- All missile tests involving a solid propellant will be assumed to have provided information with which the KPA can apply to make progress on solid-fuel ICBMs;

- All missiles that are believed to be nuclear-capable and cannot range the continental United States will be assumed to contribute to North Korea's quest for mastery over battlefield nuclear use; and

- All missiles that are believed to be MRV-capable will be assumed to contribute to MRV mastery, even if the missiles were equipped with only one reentry vehicle during testing.

With these assumptions, the 42 tests since January 2021 can be coded, as shown in Appendices A and B below.

DISCUSSION

The data shows that much of North Korea's recent missile tests present varying opportunities for progress on the five capabilities listed in the Great Program. Testing results were most likely to benefit advancements in mastering TNWs and solid-fuel ICBMs. On the other hand, MRVs received less significant attention, while even less progress was observed for submarine strategic forces and HGVs.

The timing of the tests should also be examined further. At initial glance, the timeline points to a North Korea that seemed uninterested in testing its intercontinental and multiple-warhead capabilities until early 2022, nearly a year after The Great Program was read aloud at the Worker's Party Congress, which has led some—especially in South Korea—to believe that the ICBM tests were part of a broader message that Kim wished to send to then-incoming President Yoon Suk-yeol.[45] Others speculate that

42   Caitlin Talmadge, "What Putin's Nuclear Threats Mean for the U.S.," *Wall Street Journal*, March 3, 2022, https://www.wsj.com/articles/what-putins-nuclear-threats-mean-for-the-u-s-11646329125?mod=article_relatedinline.

43   Van Diepen, "Another North Korean 'Hypersonic' Missile?"

44   Self-evident indicators (e.g., "hypersonic missile" tests contribute to HGVs) are omitted from the methodology.

45   Khang Vu, "Why Now? The Timing of North Korea's ICBM Test," Lowy Institute, March 30, 2022, https://www.lowyinstitute.org/the-interpreter/why-now-timing-north-korea-s-icbm-test.

the tests were a product of North Korean opportunism as a result of the chaos generated by Russia's invasion of Ukraine.[46] However, recent satellite imagery analysis of the Sohae Satellite Launching Station by CSIS date the preparations for ICBM testing back to 2019 following the collapse of talks at Hanoi, indicating that North Korean efforts to advance its strategic nuclear capabilities occurred in tandem with its developments on TNWs, contrary to what the timeline may imply.[47]

North Korean efforts on a given capability appear directly proportional to the short- or medium-term benefits that advancements in such capabilities would bring to Pyongyang. At this point in time, none of the items in North Korea's "to-do list" can be marked as complete. The North Korean nuclear deterrent is formidable but far from robust; it is therefore advantageous for Kim to focus on capabilities that provide the most immediate output relative to input. Establishing a North Korean nuclear dyad through a strategic submarine force, while massively beneficial in theory, entails far too many major technological advancements and changes to nuclear command and control to warrant Kim's immediate attention.

On the other hand, the relatively muted response by the United States and its NATO allies to Putin's nuclear signaling over Ukraine shows that North Korea stands to reap substantial benefits in the immediate future by commanding the means to threaten nuclear first use in a conventional conflict. North Korean development of TNWs predates the Russia-Ukraine war, but if Kim had any reason to doubt the utility of investing in nuclear weapons for battlefield use, it is all but certain that this is no longer the case. Therefore, Pyongyang is likely to continue pouring its efforts into mastering battlefield nuclear use as progress on its strategic submarine forces lags.

Finally, the research yields an additional finding that does not fit within the methodology but nevertheless warrants attention: North Korean missiles are becoming more survivable and more responsive. Advancements in solid-fuel propellants, the ampoulization of existing liquid-fuel missiles, and the introduction of new means of launch all point to North Korean missiles that will be harder to track, target, and destroy. The recent changes to nuclear launch authority also serve as a message from Pyongyang to Washington and Seoul regarding its nuclear responsiveness: if there was previously any chance that a successful assassination attempt on Kim Jong-un would immediately neutralize North Korea's nuclear forces, that possibility no longer exists.

CONCLUSION

The U.S.-ROK alliance is now headed by two presidents who are unlikely to placate Kim Jong-un for their personal agendas. Yet, as Kim amasses the credibility required for successful nuclear coercion, the decoupling pressures on the U.S.-ROK alliance will only intensify. Washington may reconsider risking a nuclear attack on its own soil to defend a distant ally, while Seoul will feel pressured to respond to North Korean nuclear advancements with nuclear weapons of its own.[48]

46   Justin McCurry, "Testing Times: Why North Korea's Missile Launches Should Worry the West," *The Guardian*, March 31, 2022, https://www.theguardian.com/world/2022/mar/31/testing-times-why-north-koreas-missile-launches-should-worry-the-west.

47   Joseph S. Bermudez, Jr., Victor Cha, and Jennifer Jun, "Recent and Future Developments of the Sohae Satellite Launching Station," Beyond Parallel, CSIS, April 13, 2022, https://beyondparallel.csis.org/recent-and-future-developments-of-the-sohae-satellite-launching-station/.

48   In-bum Chun, "It's Time for South Korea to Embrace Tactical Nuclear Weapons to Defend Itself," NK News, May 3, 2022, https://www.nknews.org/pro/its-time-for-south-korea-to-embrace-tactical-nuclear-weapons-to-defend-itself/.

It Is vital that the United States and South Korea recognize the value of their alliance, and Washington should redouble efforts to dissuade Seoul from pursuing a nuclear option that is sure to invite further instability. For example, Seoul has been aggressively pursuing a homeland artillery interception system such as Israel's Iron Dome.[49] Instead, the U.S.-ROK alliance can explore other non-nuclear responses to the North Korean nuclear threat, such as deepening cooperation on conventional counterforce.[50] Future research will focus on assessing the benefits and costs of such options.

Ultimately, none of the courses of action are silver bullets for the North Korean nuclear threat. The task of securing the Korean Peninsula remains daunting, and there is much work to be done to ensure that East Asia is not at the mercy of North Korea and Kim Jong-un's nuclear aspirations.

49   Si-young Choi, "Military to Build 'Korean Iron Dome', Enhance Strike Capability," *Korea Herald*, September 2, 2021, http://www.koreaherald.com/view.php?ud=20210902000845.
50   Lauren Sukin and Toby Dalton, "Why South Korea Shouldn't Build Its Own Nuclear Bombs," War on the Rocks, October 26, 2021, https://warontherocks.com/2021/10/why-south-korea-shouldnt-build-its-own-nuclear-bombs/.

NORTH KOREAN MISSILE TESTS, 2021–PRESENT

(AS OF NOVEMBER 2022)

Date	Missile Type	Notes	Potential Relation to the Five Capabilities in the Great Program
2021-03-21	Cruise	Short-range	TNWs
2021-03-25	SRBM	KN-23	Solid-fueled ICBMs; TNWs
2021-09-13	Cruise	Intermediate-range	TNWs
2021-09-16	SRBM	KN-23, Rail-mobile	Solid-fueled ICBMs; TNWs
2021-09-29	HGV/MaRV	Hwasong-8	HGVs
2021-10-19	SLBM	Submarine-launched KN-23 variant	Solid-fueled ICBMs; TNWs; Submarine strategic forces
2022-01-04	HGV/MaRV	Hwasong-8	HGVs; TNWs
2022-01-11	HGV/MaRV	Hwasong-8	HGVs; TNWs
2022-01-13	SRBM	KN-23, Rail-mobile	Solid-fueled ICBMs; TNWs
2022-01-17	SRBM	KN-24	Solid-fueled ICBMs; TNWs
2022-01-25	Cruise		TNWs
2022-01-30	IRBM	Hwasong-12	TNWs
2022-02-27	ICBM	Hwasong-17, First stage only	Solid-fueled ICBMs; MRVs
2022-03-04	ICBM	Hwasong-17, First stage only	Solid-fueled ICBMs; MRVs
2022-03-16	ICBM	Hwasong-17, First stage only	Solid-fueled ICBMs; MRVs
2022-03-24	ICBM	Hwasong-15 (suspected)	Solid-fueled ICBMs; MRVs
2022-04-16	New-type ballistic missile		Solid-fueled ICBMs; TNWs
2022-05-04	ICBM (suspected)		Solid-fueled ICBMs; MRVs
2022-05-07	SLBM	Submarine-launched KN-23 variant	Solid-fueled ICBMs; TNWs; Submarine strategic forces

Date	Type	Name	Desired Capabilities
2022-05-12	Unidentified SRBM		Solid-fueled ICBMs; TNWs
2022-05-25	ICBM (suspected)	Hwasong-17 (suspected)	Solid-fueled ICBMs; MRVs
2022-06-05	Unidentified SRBM		Solid-fueled ICBMs; TNWs
2022-08-17	Cruise		TNWs
2022-09-25	SLBM	Submarine-launched KN-23 variant	Solid-fueled ICBMs; TNWs; Submarine strategic forces
2022-09-29	SRBM		Solid-fueled ICBMs; TNWs
2022-10-01	SRBM		Solid-fueled ICBMs; TNWs
2022-10-04	IRBM	New and unnamed	Solid-fueled ICBMs; TNWs
2022-10-06	SRBM	KN-25	Solid-fueled ICBMs; TNWs
2022-10-09	SRBM	KN-25	Solid-fueled ICBMs; TNWs
2022-10-14	SRBM		Solid-fueled ICBMs; TNWs
2022-10-28	SRBM		Solid-fueled ICBMs; TNWs
2022-11-02	Cruise		TNWs
2022-11-02	SRBM		Solid-fueled ICBMs; TNWs
2022-11-02	Surface-to-Air		
2022-11-03	SRBM		Solid-fueled ICBMs; TNWs
2022-11-03	ICBM	Hwasong-15 (suspected)	Solid-fueled ICBMs; MRVs
2022-11-04	SRBM	KN-25	Solid-fueled ICBMs; TNWs
2022-11-04	SRBM		Solid-fueled ICBMs; TNWs
2022-11-05	SRBM		Solid-fueled ICBMs; TNWs
2022-11-09	SRBM		Solid-fueled ICBMs; TNWs
2022-11-16	SRBM		Solid-fueled ICBMs; TNWs
2022-11-17	ICBM	Hwasong-17	Solid-fueled ICBMs; MRVs

Source: "North Korean Missile Launches & Nuclear Tests: 1984-Present," *Missile Threat*, CSIS, March 24, 2022, https://missilethreat.csis.org/north-korea-missile-launches-1984-present/.

Note: HGV/MaRV tests are coded in yellow. SLBM tests are coded in blue. SRBM tests are coded in green. IRBM tests are coded in orange. IBCMs are coded in red. Submarine-launched missiles are coded in blue. Other missiles were left uncolored.

Gray Zone Security Assurances

The Case of Russia's Invasion of Ukraine

By Dilan Ezgi Koç[1]

"Ukraine has already learned this lesson the hard way, and will hardly agree to be given again 'assurances' instead of firm and clear guarantees. . . . We just cannot afford to have a Budapest Memorandum 2."[2]
— Oleksandr Merezhko, April 2022

On February 24, 2022, nuclear-armed Russia started an unprovoked and unjustified war against Ukraine, a country which had given up and disavowed nuclear weapons. The Russian military launched the largest attack by land, air, and sea conducted on European soil since World War II. The fighting has already precipitated over 10,000 deaths and a refugee crisis in the European continent, with millions of Ukrainians forced to flee their homes. Moscow's decision to violate Ukrainian sovereignty and territorial integrity and challenge the state's very existence wreaks havoc on European security and the future of the global order.

In the post–Cold War era, the transatlantic community had so far been relatively unconcerned about the possibility of nuclear escalation or a large-scale war between great powers.[3] Russian president Vladimir Putin's nuclear brinkmanship shattered this complacency. From the very start of the war, the explicit and thinly veiled nuclear threats and signaling by Putin have made the nuclear dimension

---

1   Dilan Ezgi Koç is a recent graduate of Yale University's Jackson School of Global Affairs. Views represented are her own and do not represent any past or future employer. Author's note: I thank Alexandre Debs, Stephen Herzog, Andrew Reddie, the expert external reviewer, and the Project on Nuclear Issues (PONI) team for their substantial feedback and support throughout my research and writing process.

2   David Brennan, "As NATO Weighs Nuclear War Risk, U.S. Security 'Assurances' Upset Ukraine," *Newsweek*, April 27, 2022, https://www.newsweek.com/nato-nuclear-war-risk-russia-us-ukraine-1701338.

3   Tyler Bowen, "Russia's Invasion of Ukraine and NATO's Crisis of Nuclear Credibility," War on the Rocks, April 20, 2022, https://warontherocks.com/2022/04/russias-invasion-of-ukraine-and-natos-crisis-of-nuclear-credibility/.

a focal point of the conflict.[4] His statements, such as that any "who tries to stand in our way" will be met with "consequences such as you have never seen in your entire history," have raised fears that the invasion could lead to nuclear escalation, whether intentional or unintentional.[5]

Russian aggression against Ukraine is even more alarming given that Moscow invaded a country to whom it pledged security assurances. In addition to casting away several international treaties and principles, mainly the UN Charter, the Russian invasion of Ukraine is an egregious violation of the Budapest Memorandum of 1994. Signed by the United States, United Kingdom, and Russia, the Budapest Memorandum provided security assurances to Ukraine—alongside Belarus and Kazakhstan—to encourage Kyiv's accession to the Treaty on the Nonproliferation of Nuclear Weapons (NPT) as a non-nuclear weapon state (NNWS). These security assurances mainly were commitments by the three nuclear-weapon states (NWSs) to respect Ukrainian independence and territorial integrity and to refrain from threats or use of force against the state.[6] However, the memorandum did not specify a course of action to be followed or an enforcement mechanism to be used in the event of a violation or abandonment of these security assurances. Given its ambiguity, the effect and implications of these assurances meant different things for different actors. According to former U.S. diplomat Steven Pifer, who was present at the talks, "there is an obligation on the United States that flows from the Budapest Memorandum to provide assistance to Ukraine, and … that would include lethal military assistance."[7] Yet, according to others, such as officials in Kyiv, these assurances are ambiguous; they did not protect Ukraine from its aggressive neighbor, and the state has been left to fight alone.[8] As the Russian invasion continues to violate this multilateral agreement on a daily basis, the need for reevaluating the effectiveness of security assurances is more urgent than ever.

How does the strength of security assurances affect the risk of interstate conflict involving nuclear-armed great powers? Put more simply, what does it take to deter great power aggression against a third-party state? What makes a security assurance "assuring" and seen as reliable protection? And what are the roles and implications of security assurances in the current war in Ukraine? This paper takes an inductive theory-building approach to answer these questions by focusing on the invasion of Ukraine. The paper posits that variation in security assurances given from great powers to third-party states impacts the propensity that aggressors will initiate military conflict against the third party.

The paper proceeds as follows. First, it introduces the concept of security assurances and offers a review of the existing literature, highlighting some existing gaps. Consequently, the second part lays out a new theoretical typology of security assurances based on their varying strength. It also assesses the utility of security assurances for preventing great power conflict by testing relevant hypotheses and assumptions from the extant literature. These arguments and analyses are then further contextualized and examined through a case study of the ongoing war in Ukraine. The paper concludes with policy implications and further avenues for academic research.

---

4   Alexander K. Bollfrass and Stephen Herzog, "The War in Ukraine and Global Nuclear Order," *Survival* 64, no. 4 (2022), doi:10 .1080/00396338.2022.2103255.

5   Vladimir Putin, "Address by the President of the Russian Federation," (speech, Moscow, February 24, 2022), http://en.kremlin.ru/events/president/transcripts/67843.

6   "Memorandum on Security Assurances in Connection with Ukraine's Accession to the Treaty on the Non-Proliferation of Nuclear Weapons," United Nations, December 5, 1994, https://treaties.un.org/Pages/showDetails.aspx?objid=0800000280401fbb.

7   Oleksandr Zaytsev, Steven Pifer, and Robert Einhorn, "Event: The Budapest Memorandum at 20: The United States, Ukraine and Security Assurances," (public event, Brookings Institution, Washington, DC, December 9, 2014), https://www.brookings.edu/events/the-budapest-memorandum-at-20-the-united-states-ukraine-and-security-assurances/.

8   Tammy Hughes, "Ukraine President: We Have Been Left to Fight Alone Against Russia," *London Evening Standard*, February 25, 2022, https://www.standard.co.uk/news/uk/ukraine-president-volodymr-zelensky-russia-invasion-b984578.html.

UNDERSTANDING SECURITY ASSURANCES

Security assurances are promises. They convey information about actions to be taken or avoided in the future. Often, the provider of the assurance pledges to the protégé that it will protect or improve its security, or simply will not harm it.[9] Such pledges also help states anticipate which conflicts will remain dyadic and which will draw in further belligerents.[10] Even though security assurances have been an essential feature of the regime centered around the NPT since the 1970s, they also exist outside the NPT domain. For example, in the Taiwan Relations Act of 1979, the United States *assured* Taipei it "shall make available to Taiwan such defense articles and defense services in such quantity as may be necessary to enable Taiwan to maintain a sufficient self-defense capacity as determined by the President and the Congress."[11] Another example would be the Balkan Entente of 1953, signed by Greece, Turkey, and Yugoslavia, which established assurances that military assistance would be provided if one member were attacked.[12]

But security assurances are not guarantees. Even though they are diverse in type, strength, and scope, they all lack one or more components that differentiate them from a security guarantee. An example of a security guarantee would be Article 5 of the North Atlantic Treaty—the North Atlantic Treaty Organization's (NATO) founding document—which guarantees a collective response to an armed attack against an alliance member.[13] Security guarantees establish legally binding commitments and enforcement mechanisms alongside operational strategies and capabilities to carry them out.

The security assurances Ukraine has received are part of the NPT domain, given their role in encouraging the state's membership and nuclear forbearance. There is an extensive history behind these treaty-related assurances. Despite their exclusion in the text of the NPT itself, security assurances have long been a part of NPT discussions; the states parties often made agreements and commitments outside of the formal treaty architecture, often at the NPT Review Conferences. In brief, there are two main types of NPT security assurances. First, in a negative security assurance, an NWS may promise not to use or threaten to use nuclear weapons against NNWSs. Second, in a positive security assurance, an NWS pledges to provide immediate assistance to an NNWS that is the victim of an act or threat of aggression in which nuclear weapons are used.[14]

Not all security assurances are created equal. Different types and strengths of security assurances signal different information and promises to protected states and potential aggressors alike. While some assurances can deter militarized aggression, others might fail to do so, as is occurring in Ukraine. To better understand how security assurances interact with the risk of conflict between states, both variables need to be investigated further.

---

9   Jeffrey W. Knopf, ed., *Security Assurances and Nuclear Nonproliferation* (Stanford, CA: Stanford University Press, 2012), 1–13.

10   Brett Ashley Leeds, "Do Alliances Deter Aggression? The Influence of Military Alliances on the Initiation of Militarized Interstate Disputes," *American Journal of Political Science* 47, no. 3 (2003): 427–39, https://www.jstor.org/stable/3186107.

11   U.S. Congress, House, *Taiwan Relations Act*, H.R.2479, 96th Cong., 1st sess., introduced February 28, 1979, https://www.congress.gov/bill/96th-congress/house-bill/2479.

12   Mustafa Türkeş, "The Balkan Pact and Its Immediate Implications for the Balkan States, 1930-34," *Middle Eastern Studies* 30, no. 1 (1994): 123–44, https://www.jstor.org/stable/4283618.

13   "Collective Defence – Article 5," North Atlantic Treaty Organization (NATO), Last updated July 11, 2022, https://www.nato.int/cps/en/natohq/topics_110496.htm.

14   "A Declaration by the President on Security Assurances for Non-Nuclear Weapon States Parties to the Treaty on the Non-Proliferation of Nuclear Weapons," U.S. Department of State, April 5, 1995, https://1997-2001.state.gov/global/arms/factsheets/wmd/nuclear/npt/nonucwp.html.

SECURITY ASSURANCES AND CONFLICT RISK

Security assurances were given to Ukraine in the context of its NPT accession and played an important role in its process of nuclear disarmament.[15] However, this paper does not seek to explore how security assurances impact proliferation and nonproliferation. There is already an extensive body of literature dealing with this question. While some scholars argue that a credible security guarantee in the form of a defense pact from a nuclear-armed great power can stem nuclear proliferation, others disagree and assert that alliances with nuclear-armed great powers may actually increase the risk of proliferation.[16] Furthermore, other scholars argue that security assurances do not have a significant impact on the behavior of potential proliferators or NWSs at all.[17] What has not been explicitly explored, however, is how these security assurances might specifically deter or not deter militarized aggression and the risk of interstate conflict. The war in Ukraine thus offers a real-world case to theoretically build upon this body of literature.

Security assurances can be established through formal military and security alliances or emerge externally from such structures, as per the Budapest Memorandum. Yet, the literature on security alliances does not thoroughly account for these potentially "weaker" assurances and is limited in its applicability to the ongoing conflict. In the literature on alliances and conflict, there are studies that either argue that state alliances will deter aggressors and prevent conflict or that alliance commitments can provoke war.[18] Specifically, Morrow argues that alliances operate as costly signals of intentions of great powers to intervene in a conflict involving the threatened state.[19] He further explains that the credibility of these signals results from the peacetime costs alliances impose on their members. Moreover, Johnson and Joiner argue that some alliances are more effective at deterrence than others due to variation in the credibility of alliances. For an alliance to deter a potential aggressor, the latter "must believe that the alliance is likely to be honored and that it will face a multilateral effort in the event of war."[20] This literature is helpful for understanding security assurances that emerged out of formal security alliances and their roles in conflict. Still, a gap remains in the scholarship when exploring assurances that do not fit into this category of formal alliances.

---

15 Mariana Budjeryn, "The Power of the NPT: International Norms and Ukraine's Nuclear Disarmament," *The Nonproliferation Review* 22, no. 2 (2015): 203–37, doi:10.1080/10736700.2015.1119968; Mariana Budjeryn, "Non-Proliferation and State Succession: The Demise of the USSR and the Nuclear Aftermath in Belarus, Kazakhstan, and Ukraine," *Journal of Cold War Studies* 24, no. 2 (2022): 46–94, doi:10.1162/jcws_a_01077; and Mariana Budjeryn, *Inheriting the Bomb: The Collapse of the USSR and the Nuclear Disarmament of Ukraine* (Baltimore, MD: Johns Hopkins University Press, 2022).

16 Sonali Singh and Christopher R. Way, "The Correlates of Nuclear Proliferation: A Quantitative Test," *Journal of Conflict Resolution* 48, no. 6 (2004): 859–85, https://www.jstor.org/stable/4149798; Dong-Joon Jo and Erik Gartzke, "Determinants of Nuclear Weapons Proliferation," *Journal of Conflict Resolution* 51, no. 1 (2007): 167–94, https://www.jstor.org/stable/27638542; Matthew Kroenig, "Importing the Bomb: Sensitive Nuclear Assistance and Nuclear Proliferation," *Journal of Conflict Resolution* 53, no. 2 (2009): 161–80, https://www.jstor.org/stable/20684580; Alexandre Debs and Nuno P. Monteiro, *Nuclear Politics: The Strategic Causes of Proliferation* (Cambridge: Cambridge University Press, 2017); Nuno P. Monteiro and Alexandre Debs, "The Strategic Logic of Nuclear Proliferation," *International Security* 39, no. 2 (2014): 7–51, doi:10.1162/ISEC_a_00177; Lauren Sukin, "Credible Nuclear Security Commitments Can Backfire: Explaining Domestic Support for Nuclear Weapons Acquisition in South Korea," *Journal of Conflict Resolution* 64, no. 6 (2020): 1011–42, doi:10.1177/0022002719888689; and Matthew Fuhrmann, "Spreading Temptation: Proliferation and Peaceful Nuclear Cooperation Agreements," *International Security* 34, no. 1 (2009): 7–41, doi:10.1162/isec.2009.34.1.7.

17 Joseph Pilat, "Reassessing Security Assurances in a Unipolar World," *Washington Quarterly* 28, no. 2 (2010): 159–70, https://muse.jhu.edu/article/179807.

18 Hans J. Morgenthau, *Politics Among Nations: The Struggle for Power and Peace* (New York: A.A. Knopf, 1948); Kenneth N. Waltz, *Theory of International Politics* (New York: McGraw Hill, 1979); and Thomas J. Christensen and Jack Snyder, "Chain Gangs and Passed Bucks: Predicting Alliance Patterns in Multipolarity," *International Organization* 44, no. 2 (1990): 137–68, https://www.jstor.org/stable/2706792; and John A. Vasquez, *The War Puzzle* (New York: Cambridge University Press, 1993).

19 James D. Morrow, "Alliances, Credibility, and Peacetime Costs," *Journal of Conflict Resolution* 38, no. 2 (1994): 293, https://www.jstor.org/stable/174296.

20 Jesse C. Johnson and Stephen Joiner, "Power Changes, Alliance Credibility, and Extended Deterrence," *Conflict Management and Peace Science* 38, no. 2 (2021): 178–99, doi:10.1177/0738894218824735.

More notably, Leeds hypothesizes that potential challengers are less likely to initiate a militarized dispute against a potential target with allies that are committed to come to its aid.[21] She defines alliances as "written agreements, signed by official representatives of at least two independent states, that include promises to aid a partner in the event of military conflict, to remain neutral in the event of conflict, to refrain from military conflict with one another, or to consult/cooperate in the event of international crises that create a potential for military conflict."[22] In these written alliance treaties, conditions under which promises will be operationalized are explicitly stated, in turn affecting the probability of conflict.

There appear to be some missing cases here which provide fertile grounds for investigation. For example, what about written agreements that do not establish formal state alliances but instead provide security assurances to a recipient state? Moreover, what would happen if the potential challenger were party to the written agreement? According to Leeds's definition, it could be argued that multilateral political agreements, such as the Budapest Memorandum, should be able to deter potential challengers. After all, it is a written agreement including promises to the recipient country, and the recipient has one or more allies signaling a commitment to intervene under specific conditions. This very puzzle motivates this paper, which builds on these literatures by identifying a gap: the gray zone case of the relationship between security assurances (independent variable) and interstate conflict (dependent variable).

WHAT MAKES SECURITY ASSURANCES "ASSURING?"

To address this gap in the literature, the paper first introduces a theoretical typology of security assurances, which highlights their potentially dramatic variances. It then details a theory-building exercise asking how this varying strength may affect deterrence of aggressors and the propensity of interstate conflict. It then argues that security assurances containing costly signals are more likely to deter potential aggressors and prevent conflict between states than those which lack them. To qualify as a costly signal, the security assurance must have two main components: an explicit willingness *and* capability of the patron to enforce its assurance commitments. In this context, an explicit willingness could emerge out of an act or pledge, a political or legal document, or discourse that provides a publicly stated commitment from one party to another. The explicit capability encompasses the activities and resources assigned to carry out the security commitment.

Among many other military tools, one common practice for displaying explicit operational capabilities is to deploy troops on the recipient's territory. In his seminal work, *Arms and Influence*, Schelling argues that the presence of forward-deployed "tripwire" forces on the potential target's land would create more credible threats and deter the potential aggressor. Hence, this constitutes a costly signal. This is mainly because those tripwire forces, "can die heroically, dramatically, and in a manner that guarantees that the action cannot stop there."[23] Furthermore, Reiter and Poast argue that since deterrence is more than mere signaling, in order to deter, troop deployments must actually be "sufficiently substantial to shift the local balance of power."[24]

21   Leeds, "Do Alliances Deter Aggression?"

22   Brett Ashley Leeds, Jeffrey M. Ritter, Sara McLaughlin Mitchell, and Andrew G. Long, "Alliance Treaty Obligations and Provisions, 1815–1944," *International Interactions* 28, no. 3 (2002): 237–60, doi:10.1080/03050620213653.

23   Thomas Schelling, *Arms and Influence* (New Haven, CT: Yale University Press, 1966), 22. See also Rose McDermott, Anthony C. Lopez, and Peter K. Hatemi, "'Blunt Not the Heart, Enrage It': The Psychology of Revenge and Deterrence," *Texas National Security Review* 1, no. 1 (2017): 68–89, http://hdl.handle.net/2152/63934.

24   Dan Reiter and Paul Poast, "The Truth About Tripwires: Why Small Force Deployments Do Not Deter Aggression," *Texas*

Figure 1: Theoretical Typology of Security Assurances

Strength of Assurance	Joint Defense Planning	Operational Capabilities
Strong	Yes	Yes
Weak	Yes	No
Weak	No	Yes

Source: Author's research and analysis.

Based on these assumptions, this research identifies three types of security assurances: strong security assurances and two variants of weak security assurances (see Figure 1).[25] This typology is based on two necessary conditions for security assurances to deter: joint defense planning (willingness) and operational capabilities (capability).[26] Strong security assurances include both willingness and capability. Weak security assurances include either willingness or capability but lack the other.

This paper acknowledges that these two variables might have considerable internal variation. For example, joint defense planning could come in the form of a legally binding alliance treaty (e.g., the North Atlantic Treaty) or a politically binding defense pact (e.g., the AUKUS security pact between Australia, the United Kingdom, and the United States). Operational capabilities might take the shape of a joint military command structure, troop deployments, or joint military operations. The variation within these variables may lead to an infinite number of types of security assurances, as each case could warrant a new type to understand and explain its specificities. However, for the purposes of this research and to present a parsimonious typology, this paper focuses on a selected few variables and types that have the most weight in the calculus of security assurance effectiveness.[27] Typological theorizing of security assurances not only allows for a more nuanced understanding and explanation of conflict dynamics but also serves as a practical tool for policymakers to quickly identify and diagnose the situation at hand.

STRONG AND WEAK SECURITY ASSURANCES

Strong security assurances, by their very nature, include joint defense planning and operational capabilities. While they are not credible formal alliances, they have many similarities. Hence, they provide explicit willingness and capability to carry out the commitment offered to the recipient of the assurance. This paper argues that these types of assurances are more likely to create costly signals and deter aggression. The argument follows from two main assumptions. First, joint defense planning and operational capabilities are costly signals, as they create costs for the assurance

---

National Security Review 4, no. 3 (2021): 34, doi:10.26153/tsw/13989. For other recent work expressing skepticism about the utility of small tripwire forces see, Matthew Fuhrmann, "On Extended Nuclear Deterrence," *Diplomacy & Statecraft* 29, no. 1 (2018): 51–73, doi:10.1080/09592296.2017.1420526; Brian Blankenship and Erik Lin-Greenberg, "Trivial Tripwires? Military Capabilities and Alliance Reassurance," *Security Studies* 31, no. 1 (2022): 92–117, doi:10.1080/09636412.2022.2038662; and David M. Allison, Stephen Herzog, and Jiyoung Ko, "Under the Umbrella: Nuclear Crises, Extended Deterrence, and Public Opinion," *Journal of Conflict Resolution* 66, no. 10 (May 2022), doi:10.1177/0022002722110025.

25   It is also possible to include a "no security assurances" category, which includes neither of the two necessary conditions. This, however, is not part of the scope of this research.

26   In this respect, the theory presented here draws on the principles of the "willingness" and "opportunity" framework laid out in Debs and Monteiro, *Nuclear Politics*.

27   Alexander L. George and Andrew Bennett, *Case Studies and Theory Development in the Social Sciences* (Cambridge, MA: MIT Press, 2005), 164–86.

provider. Second, costly signals are more likely to deter because the potential aggressor deems them to be more credible than rhetorical pledges alone.[28] A threat or an assurance becomes credible if the act of sending it creates a cost for the sender it would not be willing to incur otherwise.[29]

These types of signals are costly for the sender for two reasons. First, these signals are costly to send—"ex ante costs." Joint defense planning, strategy, troop mobilization, and joint operations entail significant costs. A party that is less committed to its assurance should be less likely to incur these costs.[30] Second, these signals are costly in the event that the patron fails to honor them—"ex post costs." The provider of the assurance can suffer costs if it violates or abandons the assurances via long-term damage to its reputation, domestic political consequences and audience costs, and a greater risk of war or "entrapment" in unwanted conflicts.[31] Similar to deterrent threats, the purpose of providing a security assurance is to avoid having to execute it.[32] The strongest and most effective security assurances are thus the ones that deter aggression and are never tested.

To illustrate, the Quadrilateral Security Dialogue possesses the components to qualify as a strong security assurance. It was established by Australia, India, Japan, and the United States during a time of crisis in 2004 in the aftermath of the Indian Ocean tsunami. It then became a diplomatic dialogue in 2007. Quad heads of state have expressed their explicit willingness to uphold their common goal of security and cooperation. They have committed to "a shared vision for an Indo-Pacific region that is free, open, resilient and inclusive," and to cooperation in areas ranging from cybersecurity and emerging technologies to climate change.[33] Furthermore, the members of the Quad have established a basis for regular defense cooperation by conducting naval exercises and sharing military logistics and intelligence.[34] This includes the previously trilateral naval exercise, Malabar, which has been expanded to include Australia. This maritime cooperation is intended to improve the planning, training, and employment of combat tactics between the participating nations.[35] Having both joint defense planning between members and coordinated operational capabilities to carry them out endows the Quad with costly signaling mechanisms that increase the members' ability to deter aggression.

By contrast, weak security assurances represent a gray zone. They are somewhere on the spectrum between a strong security assurance and a recipient state having no security assurances or simply

---

28   James D. Fearon, "Signaling Versus the Balance of Power and Interests: An Empirical Test of a Crisis Bargaining Model," *Journal of Conflict Resolution* 38, no. 2 (1994): 236–69, https://www.jstor.org/stable/174295; and James D. Fearon, "Signaling Foreign Policy Interests: Tying Hands versus Sinking Costs," *Journal of Conflict Resolution* 41, no. 1 (1997): 68–90, https://www.jstor.org/stable/174487.

29   Ibid.

30   James D. Morrow, "Alliances: Why Write Them Down?," *Annual Review of Political Science* 3 (2000): 63–83, doi:10.1146/annurev.polisci.3.1.63.

31   Alastair Smith, "Extended Deterrence and Alliance Formation," *International Interactions* 24, no. 4, (1998): 315–43, doi:10.1080/03050629808434934; James D. Fearon, "Domestic Political Audiences and the Escalation of International Disputes," *American Political Science Review* 88, no. 3 (1994): 577–92, https://www.jstor.org/stable/2944796; and Glenn H. Snyder, "The Security Dilemma in Alliance Politics," *World Politics* 36, no. 4 (1984): 461–95, doi:10.2307/2010183.

32   Matthew Fuhrmann and Todd S. Sechser, "Signaling Alliance Commitments: Hand-Tying and Sunk Costs in Extended Nuclear Deterrence," *American Journal of Political Science* 58, no. 4 (2014): 919–35, https://www.jstor.org/stable/24363534; and Todd S. Sechser and Mathew Fuhrmann, *Nuclear Weapons and Coercive Diplomacy* (Cambridge: Cambridge University Press, 2017).

33   Joe Biden, Narendra Modi, Scott Morrison, and Yoshihide Suga, "Our Four Nations are Committed to a Free, Open, Secure and Prosperous Indo-Pacific Region," *Washington Post*, March 13, 2021, https://www.washingtonpost.com/opinions/2021/03/13/biden-modi-morrison-suga-quad-nations-indo-pacific/.

34   Sumitha Narayanan Kutty and Rajesh Basrur, "The Quad: What It Is – And What It Is Not," *The Diplomat*, March 24, 2021, https://thediplomat.com/2021/03/the-quad-what-it-is-and-what-it-is-not/.

35   "Second Phase of Exercise Malabar 2021 Set to Commence," *Naval Technology*, October 11, 2021, https://www.naval-technology.com/news/exercise-malabar-2021-second-phase/.

verbal pledges. They signal some level of involvement, promises, and commitments by the provider of the assurance and affect state behavior. Yet, they are much more ambiguous and less firm than strong security assurances and do not carry accompanying costly signals. They either lack joint defense planning or operational capabilities. These types of assurances are thus more likely to be challenged and less likely to deter aggression. To demonstrate, the Budapest Memorandum of 1994 provides an emblematic case of a weak security assurance. On the one hand, the memorandum's nuclear-armed signatories made several promises to Ukraine, some of which are established under international law. On the other hand, the document does not provide any specific plan or strategy for how to operationalize these promises, and it lacks operational capabilities that would enhance its credibility.

CASE: THE RUSSIAN INVASION OF UKRAINE

The Budapest Memorandum and its implications as a weak security assurance deserve significant attention due to the current and future policy relevance of the ongoing war in Ukraine. This section analyzes how the memorandum came to be, what it promises, and how it has shaped the war. To begin, Ukraine's path to signing the memorandum must be understood.

UKRAINE'S NUCLEAR INHERITANCE

With the dissolution of the Soviet Union in 1991, Ukraine became an independent successor state and host to what was then the world's third-largest nuclear arsenal. According to several credible estimations, the inherited Soviet nuclear arsenal consisted of around 2,000 strategic warheads, 176 intercontinental ballistic missiles (ICBMs), 44 strategic bombers, and 2,600 tactical nuclear weapons.[36] By 1996, all nuclear munitions had been transferred to Russia, and Ukraine officially became an NNWS under the NPT alongside Belarus and Kazakhstan.

Even though Kyiv's disarmament is considered one of the greatest achievements of the NPT and the nonproliferation regime, the Ukrainian path to denuclearization was rocky at best. When it was still under Soviet rule, Ukraine declared its intention of becoming a permanently neutral state that would not accept, produce, or purchase nuclear weapons, as per its Declaration of State Sovereignty.[37] However, following its independence, Ukraine instead used its nuclear arsenal as a bargaining chip.[38] Kyiv sought to achieve several objectives in parting ways with its inherited arsenal and becoming an NNWS party to the NPT. In addition to its pursuit of becoming a member of the international community in good standing, Ukraine wanted technical and financial assistance to eliminate the post-Soviet nuclear infrastructure it inherited on its land. Kyiv also demanded low-enriched uranium (LEU) supplies to be used in fuel rods in its nuclear power reactors in compensation for the highly enriched uranium (HEU) that its strategic nuclear warheads had contained.[39]

Though the above were all crucial factors in Ukrainian decisionmaking, Kyiv's main focus was on its security concerns.[40] Ukraine wanted to receive legally binding security guarantees, mainly from

36   Budjeryn, "The Power of the NPT."

37   "Declaration of State Sovereignty of Ukraine," Government of Ukraine, July 16, 1990, http://static.rada.gov.ua/site/postano-va_eng/Declaration_of_State_Sovereignty_of_Ukraine_rev1.htm.

38   Mariana Budjeryn, "The Breach: Ukraine's Territorial Integrity and the Budapest Memorandum," Woodrow Wilson Center Nuclear Proliferation International History Project, *Issue Brief*, no. 3 (2014), https://www.wilsoncenter.org/sites/default/files/media/documents/publication/Issue%20Brief%20No%203--The%20Breach--Final4.pdf.

39   Steven Pifer, *The Trilateral Process: The United States, Ukraine, Russia and Nuclear Weapons* (Washington, DC: Brookings, May 2011), https://www.brookings.edu/research/the-trilateral-process-the-united-states-ukraine-russia-and-nuclear-weapons/.

40   Mariana Budjeryn and Matthew Bunn, *Budapest Memorandum at 25: Between Past and Future* (Cambridge, MA: Project on Managing the Atom, Harvard Belfer Center, March 2020), https://www.belfercenter.org/publication/budapest-memoran-dum-25-between-past-and-future.

the United States, to ensure its sovereignty and territorial integrity and replace the possible security benefits it could derive from having nuclear weapons.[41] Hoping to achieve this goal, the Ukrainian government drafted a treaty. This treaty was envisioned to be legally binding, signed by all NPT NWSs, and would have guaranteed Kyiv's national security as a result of its accession to the NPT.[42] When the NWSs rejected the draft treaty, Ukraine was instead forced to settle for weak security assurances.

The Ukrainian nuclear question and Kyiv's accession to the NPT were priorities for both the first Bush and the subsequent Clinton administrations as well as for Russia. They feared that nuclear weapons scattered among the post-Soviet states could lead to an increase in the number of nuclear-armed states.[43] Ensuring that these states would return their inherited nuclear arsenals to Russia and become parties to the NPT was a foreign policy priority for Washington and Moscow. However, neither the United States nor any other NWS was willing to provide the security commitments that Ukraine demanded.[44] Following difficult negotiations between Russia, the United States, the United Kingdom, and Ukraine, Kyiv agreed to a somewhat "watered-down version" of its demands from the NWSs, resulting in the now infamous "Memorandum on Security Assurances in Connection with Ukraine's Accession to the Treaty on the Non-Proliferation of Nuclear Weapons," popularly known as the Budapest Memorandum. Similarly, the former Soviet Republics of Kazakhstan and Belarus also signed their respective Budapest Memorandums in the leadup to their NPT accession as NNWSs.

THE BUDAPEST MEMORANDUM AND WEAK SECURITY ASSURANCES

The Budapest Memorandum is a multilateral agreement signed on December 5, 1994, by Ukraine, Russia, the United Kingdom, and the United States. It made political commitments to Ukraine in exchange for welcoming Kyiv's accession to the NPT as an NNWS and the removal of all nuclear weapons from the state's sovereign territory.[45] In just six paragraphs, the Budapest Memorandum— even though it is not a legally binding treaty—reiterated several clauses and agreements established by other international law documents, norms, and treaties. These principles and commitments were contained in the UN Charter, the Helsinki Final Act of the Organization for Security and Cooperation in Europe (OSCE), and various diplomatic discussions of NPT-related nuclear security assurances.

The Budapest Memorandum promised Ukraine weak security assurances. The signatories to the document certainly established a willingness to assure Kyiv. They reaffirmed their commitment to respect the independence, sovereignty, and existing borders of Ukraine and refrain from coercing it economically, all in accordance with the OSCE Final Act. They also reaffirmed their obligation to refrain from the threat or use of force against the territorial integrity or political independence of Ukraine, in accordance with the UN Charter. Furthermore, the signatories provided their commitment not to use nuclear weapons against any NPT NNWSs except in the case of self-defense and to seek immediate UN Security Council action if Ukraine should become a victim of aggression involving

---

41   Pifer, *The Trilateral Process*.

42   "Letter from Foreign Minister A. Zlenko to President L. Kravchuk on Draft Treaty on Security Guarantees," Woodrow Wilson Center History and Public Policy Program Digital Archive, Central State Archive of Ukraine, collection 5233, catalogue 1, File 280, June 3, 1993, obtained by Mariana Budjeryn and translated by Volodymyr Valkov, https://digitalarchive.wilsoncenter.org/document/119818.

43   Steven Pifer, "Why Care About Ukraine and the Budapest Memorandum," Brookings Institution, December 5, 2019, https://www.brookings.edu/blog/order-from-chaos/2019/12/05/why-care-about-ukraine-and-the-budapest-memorandum/.

44   Budjeryn, "The Power of the NPT"; and Budjeryn and Bunn, Budapest Memorandum at 25.

45   Political commitments are not legally binding since international law provides no normative backing force for them. See, for example, Nick Flynn and Nicola Peart, "The Role of Political Agreement in a Legally Binding Outcome," European Capacity Building Initiative, November 2010, https://oxfordclimatepolicy.org/sites/default/files/TheRoleofPoliticalAgreementInALegallyBindingOutcome.pdf.

nuclear weapons.[46] And lastly, the signatories agreed to consult in the event of a situation concerning the outlined commitments.

However, willingness is just one of the two necessary conditions for a strong security assurance; the Budapest Memorandum represents a weak assurance because it lacked the requisite operational capability to honor this willingness. Kyiv's occasional joint military exercises with the United States and other NATO troops along with capacity-building activities for the Ukrainian Armed Forces were indeed crucial for improving its military capabilities.[47] However, the memorandum did not establish an operational capability in any form to increase the credibility of its assurances. And hence, it did not involve costly signals to enhance the credibility of deterrence. Lacking a plan of action, enforcement mechanisms, and operational capabilities to carry out its commitments, the Budapest Memorandum left enough leeway for Russia to take a chance by testing its credibility.

The memorandum's consultation clause was invoked for the first time in 2014 following the Russian invasion of Crimea. By arguing that "the security assurances were given to the legitimate government of Ukraine but not to the forces that came to power following the coup d'état in Kyiv," Russian foreign minister Sergey Lavrov declined to participate in the consultations.[48] This rhetorical evasion helped facilitate Russian belligerence, resulting in the annexation of Ukraine's Crimean Peninsula and a years-long conflict that continues in the Donbas region.[49] According to a 2019 report by the Office of the UN High Commissioner for Human Rights, by that point the conflict in eastern Ukraine had led to at least 3,339 civilian deaths, damage to civilian infrastructure and schools, limits on freedom of movement across the contact line, and systemic violation of the rights, liberty, and security of Ukrainian citizens.[50] That situation is exponentially worse today due to Russia's large-scale 2022 invasion. Without a UN Security Council mandate to intervene and a reasonable case to be made for Russian self-defense, Moscow cast away its memorandum commitment to respect the independence, sovereignty, and territorial integrity of Ukraine. The annexation of Crimea and the activities of the "little green men"—Russian forces operating without insignia—in Ukraine violated several assurances provided in the Budapest Memorandum.

PUTIN'S WAR OF AGGRESSION

In an address on state television on February 24, 2022, Putin announced a "special military operation" against Ukraine, aiming "to protect people who have been subjected to bullying and genocide . . . for the last eight years. And for this we will strive for the demilitarization and denazification of Ukraine."[51] He followed the address by initiating a full-scale invasion of Ukraine by land, air, and sea.

46   "Memorandum on Security Assurances in Connection with Ukraine's Accession to the Treaty on the Non-Proliferation of Nuclear Weapons," United Nations.

47   "Ukraine Holds Military Drills with U.S. Forces, NATO Allies," Reuters, September 20, 2021, https://www.reuters.com/business/aerospace-defense/ukraine-holds-military-drills-with-us-forces-nato-allies-2021-09-20/.

48   David S. Yost, "The Budapest Memorandum and Russia's Intervention in Ukraine," *International Affairs* 91, no. 3 (2015): 505–38, https://www.jstor.org/stable/24539145; and "Ukraine, Nuclear Weapons, and Security Assurances at a Glance," Arms Control Association, February 2022, https://www.armscontrol.org/factsheets/Ukraine-Nuclear-Weapons.

49   Denis Corboy, William Courtney, and Kenneth Yalowitz, "Hitting the Pause Button: The 'Frozen Conflict' Dilemma in Ukraine," *The National Interest*, November 6, 2014, https://nationalinterest.org/feature/hitting-the-pause-button-the-frozen-conflict-dilemma-ukraine-11618?nopaging=1.

50   Office of the United Nations High Commissioner for Human Rights, *Report on The Human Rights Situation in Ukraine 16 May to 15 August 2019* (Geneva: 2019), https://www.ohchr.org/sites/default/files/Documents/Countries/UA/ReportUkraine-16May-15Aug2019_EN.pdf.

51   Andrew Osborn and Polina Nikolskaya, "Russia's Putin Authorizes 'Special Military Operation' Against Ukraine," Reuters, February 24, 2022, https://www.reuters.com/world/europe/russias-putin-authorises-military-operations-donbass-domestic-media-2022-02-24/; Max Fisher, "Word by Word and Between the Lines: A Close Look at Putin's Speech," *New York Times*, February 23, 2022, https://www.nytimes.com/2022/02/23/world/europe/putin-speech-russia-ukraine.html; and

Only four days into the invasion, Putin ordered Russia's nuclear forces into "special combat readiness," a heightened and escalatory status reminiscent of some of the nuclear crises of the Cold War.[52] Russia's decision to invade Ukraine and reject the state's very existence once again cast down the credibility of the security assurances made to Ukraine, only this time Moscow's actions have been more egregious and will carry longer-term effects.[53]

A pretext for Putin's war was Ukraine's potential membership in NATO, although Kyiv's weak security assurances were nothing like those of Article 5 security guarantees. Taking this possibility as a direct threat to Russia, Putin stated, "The Ukrainian army is waiting to get into NATO. . . . The West has explored the territory of Ukraine as a future theater, future battlefield, that is aimed against Russia."[54] Kyiv's aspirations for NATO membership are news to neither NATO nor Russia. Especially since the Russian annexation of Crimea, joining NATO has become one of the top priorities for Kyiv. President Volodymyr Zelenskyy has pleaded for his country to be brought into NATO and the European Union for years.[55] However, the Ukrainian president has grown frustrated about not receiving a timeline or a firm commitment of membership.

Just one month before the invasion, the Biden administration indicated it had no immediate plans to bring Ukraine into the alliance.[56] Yet, citing NATO's "open-door policy" based on Article 10 of the North Atlantic Treaty, U.S. president Joe Biden stated that each sovereign country has the right to decide for itself whether it wants to join the alliance.[57] Membership in NATO would include Ukraine within the alliance's collective defense infrastructure and provide security guarantees to Kyiv. These security guarantees would be more likely to create costly signals and deter Russian aggression, but they would also result in costly signaling for NATO and taking on risks vis-à-vis Russia. Moscow made it clear that even a possibility of Ukraine joining NATO would threaten Russian national security. In March 2022, Zelensky conceded, "It is clear that Ukraine is not a member of NATO; we understand this. . . . For years we heard about the apparently open door, but have already also heard that we will not enter there, and these are truths and must be acknowledged."[58] Ukraine's future within or outside of NATO remains unclear, but for the moment Kyiv must settle for weak security assurances, rather than strong security guarantees, that do not appear to deter.

By starting an unprovoked and unjustified war against Ukraine, Russia violated Article 2(4) of the UN Charter, which requires member states to refrain from "the use of force against the territorial integrity

---

Max Fisher, "Putin's Case for War, Annotated," *New York Times*, February 24, 2022, https://www.nytimes.com/2022/02/24/world/europe/putin-ukraine-speech.html.

52 Yuras Karmanau, Jim Heintz, Vladimir Isachenkov, and Dasha Litvinova, "Putin Puts Nuclear Forces on High Alert, Escalating Tensions," Associated Press, February 27, 2022, https://apnews.com/article/russia-ukraine-kyiv-business-europe-moscow-2e4e1cf784f22b6afbe5a2f936725550; and Christine Amanpour (@amanpour), Twitter post, March 22, 2022, 12:04 pm, https://twitter.com/amanpour/status/1506346172977491978.

53 "Ukraine Crisis: Russia Orders Troops Into Rebel-held Regions," BBC News, February 22, 2022, https://www.bbc.com/news/world-europe-60468237; and Andrew E. Kramer, "Russian Troop Movements and Talk of Intervention Cause Jitters in Ukraine," *New York Times*, April 30, 2021, https://www.nytimes.com/2021/04/09/world/europe/russia-ukraine-war-troops-intervention.html.

54 Fisher, "Word by Word and Between the Lines."

55 "Ukraine Reaffirms Desire to Join NATO After Envoy Comments," Deutsche Welle, February 14, 2022, https://www.dw.com/en/ukraine-reaffirms-desire-to-join-nato-after-envoy-comments/a-60768298.

56 Edward Wong and Lara Jakes, "NATO Won't Let Ukraine Join Soon. Here's Why," *New York Times*, January 13, 2022, https://www.nytimes.com/2022/01/13/us/politics/nato-ukraine.html.

57 "Fact Sheet: NATO Enlargement & Open Door," NATO, July 2016, https://www.nato.int/nato_static_fl2014/assets/pdf/pdf_2016_07/20160627_1607-factsheet-enlargement-eng.pdf.

58 Maite Fernandez Simon, "Ukraine's Zelensky Asks European Leaders to Do More but Admits NATO Membership is Not in the Cards," *Washington Post*, March 15, 2022, https://www.washingtonpost.com/world/2022/03/15/zelensky-remarks-ukraine-nato-membership-russian-invasion/.

or political independence of any state."[59] In addition to disregarding core tenets of international and humanitarian law, Russia cast away five of six clauses of the Budapest Memorandum.[60] In brief, Russia does not respect the independence, sovereignty, and borders of Ukraine. The statements from Putin, such as "Modern Ukraine was entirely created by Russia, more precisely, Bolshevik, communist Russia" and "Ukraine never had a tradition of genuine statehood," clearly demonstrate his hostility toward the existence of the Ukrainian state.[61] This rhetoric employs both the threat and use of force against the territorial integrity and political independence of Ukraine. And it does so by using weapons without a justified cause, such as self-defense. Without joint defense planning and any backing via operational capabilities, there was insufficient willingness to defend Ukraine and deter Russia among Kyiv's Western patrons.

Today, Ukrainian suffering continues with no end in sight. While Russian missiles and artillery are razing Ukrainian cities, the Kremlin's military forces are violating the international laws of war and basic human rights of civilians in occupied regions of Ukraine. According to Hugh Williamson, director of Human Rights Watch Europe and Central Asia, "The cases we documented amount to unspeakable, deliberate cruelty and violence against Ukrainian civilians," and "rape, murder, and other violent acts against people in the Russian forces' custody should be investigated as war crimes."[62]

The implications of the Russian invasion become even more pressing given the role of nuclear weapons in the conflict. As an NWS state party to the NPT and Budapest Memorandum, Russia committed to seek immediate UN Security Council (UNSC) action to provide assistance to Ukraine if it should "become a victim of an act of aggression or an object of a threat of aggression in which nuclear weapons are used."[63] However, Russia, as a permanent member of the UNSC, was quick to veto a UNSC Resolution that would have demanded that Moscow "immediately stop its attack on Ukraine and withdraw all troops."[64] Clearly, the Budapest Memorandum lacked the "teeth" to enforce its provisions.

NUCLEAR DYNAMICS

Since launching its full-scale invasion of Ukraine, Moscow has struck a major blow to global nuclear governance. Even though the war is between nuclear-armed Russia and non-nuclear-armed Ukraine and no use of nuclear weapons has been observed, Putin's nuclear saber-rattling put the threat of nuclear weapons at the center of the conflict. These regular nuclear threats and prospects for escalation disturbed the decades-old nuclear taboo and distorted the assumption that modern conflicts would not escalate into nuclear war. On March 14, UN secretary general Antonio Guterres even stated that "The prospect of nuclear conflict, once unthinkable, is now back within the realm of possibility."[65]

---

59   "United Nations Charter," United Nations, https://www.un.org/en/about-us/un-charter/full-text.
60   "Memorandum on Security Assurances in Connection with Ukraine's Accession to the Treaty on the Non-Proliferation of Nuclear Weapons," United Nations.
61   "Extracts from Putin's Speech on Ukraine," Reuters, February 21, 2022, https://www.reuters.com/world/europe/extracts-putins-speech-ukraine-2022-02-21/.
62   "Ukraine: Apparent War Crimes in Russia-Controlled Areas," Human Rights Watch, April 3, 2022, https://www.hrw.org/news/2022/04/03/ukraine-apparent-war-crimes-russia-controlled-areas.
63   "Memorandum on Security Assurances in Connection with Ukraine's Accession to the Treaty on the Non-Proliferation of Nuclear Weapons," United Nations.
64   "Russia Blocks Security Council Action on Ukraine," UN News, February 26, 2022, https://news.un.org/en/story/2022/02/1112802.
65   "Kremlin: Russia Would Only Use Nuclear Weapons If its Existence were Threatened," Reuters, March 22, 2022, https://www.reuters.com/world/europe/kremlin-russia-would-only-use-nuclear-weapons-if-its-existence-were-threatened-2022-03-22/.

It is true that showing off nuclear arsenals and uttering nuclear threats as a strategy is not unprecedented or unique to this conflict.[66] Many NWSs have threatened other countries with their arsenals throughout the nuclear age. For instance, in 2018, former U.S. president Donald Trump threatened North Korean supreme leader Kim Jong-un and stated that his nuclear button is "much bigger."[67] Or more recently, in April 2022, North Korean leadership stated that it would strike with nuclear weapons if South Korea were ever to attack.[68] However, what makes Putin's nuclear threats more alarming is that he is making them as he continues his assault in order to deter other states from helping Kyiv. Nuclear weapons thus help to enable his aggressive behavior.

Analysts and policymakers argue that the war is unlikely to go nuclear, but the risk is not zero.[69] Many speculate that Putin's nuclear signaling and threats are designed to control escalation, intimidate and deter the West and NATO from directly interfering in the conflict, and highlight potential risks of Western economic and military support for Kyiv.[70] Moscow has also used nuclear-capable missiles against Ukraine and sent its nuclear-armed submarines into combat drills.[71] Furthermore, in April 2022, Russia successfully tested the Sarmat ICBM, and Putin commented that this weapon will "make those who—in the heat of aggressive rhetoric—try to threaten our country think twice."[72] Kremlin officials repeatedly stated that "Russia would only use nuclear weapons if its existence were threatened."[73] It is not clear, though, whether Russia's existence and Putin's regime survival are the same.

IMPLICATIONS AND CONCLUSION

Russia's invasion of Ukraine might have changed European security dynamics for the foreseeable future. Beyond that, the implications of Russia's actions extend far beyond the borders of Ukraine and the European continent. As the costs of war continue to grow, in thousands of lives and the displacement of millions, the war also poses major challenges to the global security structure and international norms. By attacking a disarmed country to whom it pledged security assurances, Russia discarded its commitments under the Budapest Memorandum. It also disregarded the UN Charter and countless pieces of international humanitarian law while challenging the nonproliferation regime and the future of bilateral and multilateral nuclear arms control.

Moscow's security assurances to Ukraine had no effect on its actions and counted for little beyond the paper they were written on. The implications of this flagrant disregard of security assurances by a major nuclear-armed state could be far reaching.

---

66   See, for example, Bollfrass and Herzog, "The War in Ukraine and Global Nuclear Order."
67   "Trump to Kim: My Nuclear Button is 'Bigger and More Powerful'," BBC, January 3, 2018, https://www.bbc.com/news/world-asia-42549687.
68   Josh Smith, "North Korea Says It Will Strike With Nuclear Weapons If South Attacks," Reuters, April 5, 2022, https://www.reuters.com/world/asia-pacific/nkorea-says-it-will-strike-south-with-nuclear-weapons-if-attacked-kcna-2022-04-04/.
69   "Will Putin Use Nuclear Weapons in Ukraine? Our Experts Answer Three Burning Questions," Atlantic Council, May 10, 2022, https://www.atlanticcouncil.org/blogs/new-atlanticist/will-putin-use-nuclear-weapons-in-ukraine-our-experts-answer-three-burning-questions/.
70   Gustav Gressel, "Shadow of the Bomb: Russia's Nuclear Threats," European Council on Foreign Relations, July 7, 2022, https://ecfr.eu/article/shadow-of-the-bomb-russias-nuclear-threats/; and Stephen Blank, *Russian Nuclear Strategy in the Ukraine War: An Interim Report* (Oakton, VA: National Institute for Public Policy, June 2022), https://nipp.org/wp-content/uploads/2022/06/IS-525.pdf.
71   Bollfrass and Herzog, "The War in Ukraine and Global Nuclear Order."
72   "Sarmat: Russian Missile Will Make Enemies Think Twice, Says Putin," Al Jazeera, April 20, 2022, https://www.aljazeera.com/news/2022/4/20/new-russian-missile-will-make-enemies-think-twice-says-putin.
73   "Kremlin: Russia Would Only Use Nuclear Weapons If its Existence were Threatened," Reuters, March 22, 2022, https://www.reuters.com/world/europe/kremlin-russia-would-only-use-nuclear-weapons-if-its-existence-were-threatened-2022-03-22/.

First, as showed above, the Russian invasion of Ukraine consolidated Kyiv's recognition that the security assurances they received decades ago fall short of deterring Russian aggression. Kyiv's discontent with weak security assurances and search for something stronger are old news. The Ukrainian government has been expressing its desire to replace the Budapest Memorandum with "a more reliable diplomatic instrument" for years. Former Ukrainian president Petro Poroshenko promised his citizens in 2014 that he would use his "diplomatic experience to ensure the signature of an international agreement that would replace the Budapest Memorandum. Such agreement must provide direct and reliable guarantees of peace and security—up to military support in case of threat to territorial integrity."[74] Since the war started, Ukraine's leaders have been increasingly adamant about their demand for legally binding security guarantees as part of any deal to end the war. They argue that, absent NATO membership, the only way to deter Russian aggression is to receive "genuine security guarantees."[75] It is yet to be seen how this war will end and whether the signatories of the Budapest Memorandum and Kyiv will be able to agree on assurance that actually "assures" Ukraine while deterring further Russian aggression.

Second, Russia's actions cast doubt on the overall credibility and effectiveness of weak security assurances as diplomatic tools for nuclear nonproliferation. This could impact both aspiring proliferators, such as Iran, and also NNWSs that have received security assurances in the NPT context. In a setting where there is already great division between nuclear "haves" and "have nots," the erosion of confidence in assurances provided by NWSs could create nuclear incentives for NNWSs. These parties could seek formal security alliances with NWSs and place themselves under their nuclear umbrellas. Or they could pursue nuclear hedging and even initiate or accelerate plans to acquire their own nuclear arsenals to safeguard against aggression by a nuclear-armed belligerent.[76] Some observers argue that the Russian invasion of Ukraine already emboldened hawkish factions in Japan and South Korea who want to pursue independent nuclear arsenals.[77] Likewise, as nuclear negotiations with Iran wear on amid Russia's aggression, Tehran might conclude that assurances and promises are no longer credible and that nuclear weapons are the proper tool to deter adversaries.

Third, by starting an unjustified and unprovoked war and using its nuclear arsenal as a deterrent tool to enable escalation, Russia is damaging the global nuclear order. This order, which comprises nuclear nonproliferation treaties and arms control and risk reduction mechanisms, is challenged at the core by Russia's aggression and nuclear brinkmanship. It is true that the bilateral arms control processes between the United States and Russia were already suffering before the invasion. The U.S. withdrawal from the Intermediate-Range Nuclear Forces (INF) Treaty in 2019—in response to Moscow's violations—and from the Open Skies Treaty in 2020 left only one surviving arms control treaty between the two states.[78] The New Strategic Arms Reduction Treaty (New START), which will

74   Petro Poroshenko, "The Address of the President of Ukraine During the Ceremony of Inauguration," Government of Ukraine, June 7, 2014, http://www.president.gov.ua/en/news/30488.html.

75   David Brennan, "As NATO Weighs Nuclear War Risk, U.S. Security 'Assurances' Upset Ukraine," *Newsweek*, April 27, 2022, https://www.newsweek.com/nato-nuclear-war-risk-russia-us-ukraine-1701338.

76   Yost, "The Budapest Memorandum and Russia's Intervention in Ukraine"; Francesca Giovannini, "Negative Security Assurances After Russia's Invasion of Ukraine," *Arms Control Today* 52 no. 6 (2022): 6–11, https://www.armscontrol.org/act/2022-07/features/negative-security-assurances-after-russias-invasion-ukraine; and Bollfrass and Herzog, "The War in Ukraine and Global Nuclear Order."

77   Toby Dalton, Karl Friedhoff, and Lami Kim, *Thinking Nuclear: South Korean Attitudes on Nuclear Weapons* (Chicago: Chicago Council on Global Affairs, February 2022), https://www.thechicagocouncil.org/sites/default/files/2022-02/Korea%20Nuclear%20Report%20PDF.pdf; and Choe Sang-Hun, "In South Korea, Ukraine War Revives the Nuclear Question," *New York Times*, April 6, 2022, https://www.nytimes.com/2022/04/06/world/asia/ukraine-south-korea-nuclear-weapons.html.

78   C. Todd Lopez, "U.S. Withdraws from Intermediate-Range Nuclear Forces Treaty," U.S. Department of Defense, August 2, 2019, https://www.defense.gov/News/News-Stories/Article/Article/1924779/us-withdraws-from-intermediate-range-nu-

expire on February 4, 2026, places verifiable limits on the two countries' deployed strategic nuclear weapons.[79] Given Russia's violation of international nonproliferation norms and security assurances, the possibility that the United States and Russia will sit down at the table together and work to improve upon bilateral arms control seems bleak.

Furthermore, aspirations for a trilateral arms control framework involving China do not warrant much optimism. The power dynamics of the United States, Russia, and China in a period of intense multipolarity were already challenging the global nonproliferation regime before Putin's invasion of Ukraine.[80] The Trump administration had failed to get China to agree to even discuss a trilateral arms control treaty when they had made Beijing's participation in New START a sine qua non for renewal.[81] By destroying the credibility of its security assurances and turning itself into a global pariah, Russia might not qualify as a reliable partner for a trilateral arms control effort anytime soon.

In the future, it is possible to see more crises such as Ukraine that test the viability of weak security assurances from great powers and spark escalation risks based on uncertainty. To begin thinking about such crises before they happen, this paper demonstrated that weak security assurances are less likely to create costly signals and deter aggression. The contribution of this study is three-fold. First, it outlines a puzzling gap in the scholarly literature and demonstrates that the role of security assurances in interstate conflict is understudied and requires further exploration. Second, it proposes a new theoretical typology of security assurances. And third, it examines such security assurances by delving deeper into the war in Ukraine and the Budapest Memorandum, showing the implications of the analysis for policy. While this theory-building exercise is a step in the right direction for improving the study of nuclear crises, in order to better understand the implications of security assurances as a policy tool, future scholarship should introduce further theoretical innovations and case studies. This would greatly help to inform policymakers and to further scrutinize the existing literature and arguments, including this paper.

Unfortunately, Russia's war in Ukraine will likely continue for many more months, if not years. As the war continues, the camps disagreeing on the value of nuclear weapons are constantly evolving and growing further apart in their positions. On the one hand, some see nuclear weapons as a tool that could have helped Ukraine to deter Russia.[82] Some of these observers see the war as evidence of how much of a free hand a nuclear-armed belligerent has with invading an NNWS and as further proof of

<hr />

clear-forces-treaty; and Amy F. Woolf, "The Open Skies Treaty: Background and Issues," *Congressional Research Service*, June 7, 2021, https://sgp.fas.org/crs/nuke/IN10502.pdf.

79   "New START Treaty," U.S. Department of State, https://www.state.gov/new-start/. And for a discussion of the institutional design of New START, see Andrew W. Reddie, "Governing Insecurity: Institutional Design, Compliance, and Arms Control," Ph.D. dissertation, University of California, Berkeley, 2019, https://escholarship.org/content/qt2sn678w1/qt2sn678w1_no-Splash_3c9965fc7803f8795cc9369de3eb6b2c.pdf?t=q0nama.

80   Rebecca Davis Gibbons and Stephen Herzog, "Durable Institution Under Fire? The NPT Confronts Emerging Multipolarity," *Contemporary Security Policy* 43, no. 1 (2022): 50–79, doi:10.1080/13523260.2021.1998294; Eliza Gheorghe, "Proliferation and the Logic of the Nuclear Market," *International Security* 43, no. 4 (2019): 88–137, doi:10.1162/isec_a_00344; and Nicholas L. Miller and Tristan Volpe, "The Rise of the Autocratic Nuclear Marketplace," *Journal of Strategic Studies* (April 2022), doi:10.1080/01402390.2022.2052725.

81   Eliza Gheorghe and Dilan Ezgi Koç, "Not in The Cards: U.S.-China Arms Control in the Era of Multipolar Competition," *Turkish Policy Quarterly* 20, no. 3 (2021): 123–131, http://turkishpolicy.com/files/articlepdf/us-china-arms-control_en_9284.pdf; and David M. Allison and Stephen Herzog, "'What About China?' and the Threat to US–Russian Nuclear Arms Control," *Bulletin of the Atomic Scientists* 76, no. 4 (2020): 200–5, doi:10.1080/00963402.2020.1778370.

82   Alexa Phillips, "Russian Invasion 'Wouldn't Have Happened' If Ukraine Still Had Nuclear Weapons, Ukrainian Political Adviser Says," *Sky News*, March 4, 2022, https://news.sky.com/story/russian-invasion-wouldnt-have-happened-if-ukraine-still-had-nuclear-weapons-ukrainian-political-adviser-says-12556811.

the advantages of a "homegrown nuclear deterrent."[83] On the other hand, other commentators see this war as a stark reminder of the urgency of nuclear disarmament, given the nuclear risks showcased by Putin's actions.[84] Ultimately, sound policymaking in light of the war in Ukraine will benefit from accounting for the risks and benefits of nuclear deterrence, security assurances, and disarmament.

83   Choe Sang-Hun, "In South Korea, Ukraine War Revives the Nuclear Question," *New York Times*, April 6, 2022, https://www.nytimes.com/2022/04/06/world/asia/ukraine-south-korea-nuclear-weapons.html.
84   Sharon Squassoni, "Rethinking the Unthinkable: Ukraine Reveals the Need for Nuclear Disarmament," Bulletin of the Atomic Scientists, March 17, 2022, https://thebulletin.org/2022/03/rethinking-the-unthinkable-ukraine-reveals-the-need-for-nuclear-disarmament/.

Counting to Nuclear Zero Is Hard

How Can the World Prepare to Verify the Absence of Nuclear Weapons?

By Lansing S. Horan IV[1]

OVERVIEW

Whether nuclear weapons could be entirely eliminated is one of the fundamental questions of the modern era. Treaties and international agreements between countries continue to pressure nuclear weapon states (NWSs) to uphold their good-faith efforts toward disarmament. Arms control agreements have generally helped reduce stockpile sizes, down from a global combined Cold War peak of 70,300 nuclear weapons in 1986 to an estimated 12,700 in early 2022.[2] Political winds and public opinions are ever present and ever shifting, with several groups increasingly advocating for nuclear weapons abolition.

This research article will analyze the practicality of maintaining a "nuclear zero" policy.[3] Given a hypothetical world without nuclear weapons, the practical challenge will be how to keep it at zero.

1   Lansing S. Horan IV is a PhD graduate student in the department of Nuclear Science and Engineering at the Massachusetts Institute of Technology (MIT) in Cambridge, Massachusetts. This work is supported by the Department of Energy's NNSA Stewardship Science Graduate Fellowship (SSGF) program under Cooperative Agreement DE-NA0003960. Earlier in this paper's writing, Lansing was a nuclear engineer at the Air Force Technical Applications Center (AFTAC) at Patrick Space Force Base, Florida. The views expressed in this article are those of the author and do not necessarily reflect the official policy or position of the Department of Energy, the U.S. Air Force, the Department of Defense, the U.S. government, or other institutions of association.

2   Hans M. Kristensen, Matt Korda, and Robert Norris, "Status of World Nuclear Forces," Federation of American Scientists, February 23, 2022, https://fas.org/issues/nuclear-weapons/status-world-nuclear-forces/.

3   This paper does not intend to advocate positively for, nor cast judgement against, nuclear zero or nuclear abolition more broadly. These topics can understandably be polarizing and subject to strong opposing opinions. Whether or not the nuclear zero objective is "good" or "bad" (or rather, "better" or "worse" than the current non-zero status quo) is a decision that policymakers will need to contemplate. Instead, this paper attempts to take a more agnostic survey of the landscape, the core interest being toward discussion rather than debate. Fully zeroing the world's inventories of nuclear weapons seems to be a rather lofty goal. This paper assumes nuclear zero's aims as the premise for its hypothetical investigation.

The safeguards systems currently in place is often argued to be insufficient to verify the total global absence of nuclear weapons. What new and improved safeguards systems—such as additional monitoring strategies and technologies, legal frameworks for inspections and enforceability, and access requirements—are needed for thorough verification to be possible and to ensure that nuclear zero is sustainable?

This paper has five sections:

1. A brief historic background on nuclear weapons abolition for context and motivation;

2. An introduction to the International Atomic Energy Agency (IAEA)'s safeguards for verification;

3. A listing of the current limitations of IAEA safeguards;

4. A discussion of how nuclear zero requires a bolder safeguards regime; and

5. Proposals for more demanding safeguards to make nuclear zero more maintainable.

Several of the proposals in this paper are not politically feasible at the present time. However, the addition of some or all of these new protocols for verification might bolster the appeal of nuclear zero itsel, and therefore allow it more serious consideration. Even if nuclear zero is ultimately not achieved, an emboldened international verification regime could still permit for stronger arms control agreements, possibly reducing arsenals toward more minimum deterrence levels while lessening nuclear risks.

HISTORICAL BACKGROUND ON NUCLEAR WEAPON ABOLITION

THE TWENTIETH CENTURY'S NUCLEAR DISARMAMENT MOVEMENTS

Lawrence Wittner, an award-winning American historian, writer, and peace activist, argues that there were three major eras of global disarmament sentiment in the twentieth century, each stronger than the last.[4] From these three eras arose several international nuclear arms control treaties on the road toward disarmament.

The first nuclear weapons abolition era was in the 1940s—the beginning of the nuclear age itself—as a reaction to the United States using atomic bombs against Japan. In January 1946, the first-ever resolution from the UN General Assembly called for "the elimination from national armaments of atomic weapons and of all other major weapons adaptable to mass destruction."[5] This resolution also created the UN Atomic Energy Commission (UNAEC). In June 1946, the Truman administration proposed the Baruch Plan to the UNAEC.[6] If adopted, the Baruch Plan would have had the United States entirely decommission all of its nuclear weapons and share fission technology with the world, under the condition that all nations pledge to never develop nuclear weapons and agree to an inspection regime of monitoring, policing, and sanctions. As the Cold War began, these early abolition ideas quickly fell out of favor.

---

4   Lawrence S. Wittner, *Confronting the Bomb: A Short History of the World Nuclear Disarmament Movement* (Stanford, CA: Stanford University Press, 2009), https://www.sup.org/books/title/?id=9646; and Lawrence S. Wittner, David S. Patterson, and Stan Riveles, "Confronting the Bomb: A Short History of the World Nuclear Disarmament Movement," (public event, Wilson Center, Washington, DC, July 20, 2009), https://www.wilsoncenter.org/event/confronting-the-bomb-short-history-the-world-nuclear-disarmament-movement.

5   "Resolutions Adopted on the Reports of the First Committee," United Nations, General Assembly Resolution 1, 1946, http://undocs.org/en/A/RES/1(I).

6   U.S. Department of State, "The Acheson-Lilienthal & Baruch Plans, 1946," Department of State, n.d., https://history.state.gov/milestones/1945-1952/baruch-plans.

The second abolition era was in the late 1950s and 1960s, with public pressure mounting against atmospheric nuclear testing due to environmental concerns. The Eisenhower administration instituted a testing moratorium in 1958. In 1963, the Kennedy administration signed the first major multilateral international treaty on nuclear weapons, the Partial Test Ban Treaty (PTBT), which restricted all nuclear tests to be contained underground.[7] In 1968, the Treaty on the Non-Proliferation of Nuclear Weapons (NPT) was signed.[8] The NPT is the most comprehensive and influential international treaty on nuclear weapons, with 189 signatory member states, more than any other arms control or disarmament agreement. The NPT aims to "prevent the spread of nuclear weapons," "foster the peaceful uses of nuclear energy," and "further the goal of disarmament," identifying the IAEA as the organization for global nuclear safeguards and verification.[9]

The third abolition era occurred in the late 1970s and 1980s, at the close of U.S.-Soviet détente and a return to Cold War tensions. In the 1970s, the United States and the Soviet Union moved to reduce nuclear escalation. Key arms control treaties and bilateral agreements, such as SALT I, the Anti-Ballistic Missile Treaty, and SALT II, had the two nations mutually reduce their deployments of various ballistic missile classes.[10] With the fall of détente, the Reagan administration initially proposed a build-up of strategic nuclear weapons and the deployment of new intermediate-range missiles and neutron warheads in Europe. The "nuclear freeze" movement of the late 1970s and early 1980s led to widespread protests throughout Europe and the United States.[11] Plans for enhanced neutron-radiation weapons in Europe were cancelled. In 1987, the Intermediate-Range Nuclear Forces Treaty (INF), an arms control treaty of Reagan's "Zero Option" design, banned all Soviet and American intermediate-range missiles from Europe.[12]

The twentieth century concluded with a far-reaching multilateral treaty on nuclear weapons, the Comprehensive Nuclear-Test-Ban Treaty (CTBT) in 1996. If it ever enters into force, the CTBT "bans all nuclear explosions on Earth whether for military or for peaceful purposes."[13]

THE TWENTY-FIRST CENTURY'S RENEWED "GLOBAL ZERO" MOVEMENTS

In many ways, nuclear weapons abolition in the twenty-first century was brought into the mainstream dialogue in January 2007. This was when four prominent former U.S. statesmen—George Shultz, William Perry, Henry Kissinger, and Sam Nunn, all of whom had spent various parts of their careers involved in the United States' nuclear weapons establishment—published a widely read opinion piece, "A World Free of Nuclear Weapons," in the *Wall Street Journal*.[14] These individuals

---

7   "Treaty Banning Nuclear Tests in the Atmosphere, in Outer Space and Under Water (PTBT – Partial Test Ban Treaty)," Center for Nonproliferation Studies, signed August 5, 1963, https://www.nti.org/media/documents/ptbt_partial_test_ban_treaty.pdf.

8   "Treaty on the Non-Proliferation of Nuclear Weapons (NPT)," United Nations, signed July 1, 1968, https://www.un.org/disarmament/wmd/nuclear/npt/text/.

9   "Treaty on the Non-Proliferation of Nuclear Weapons (NPT)," International Atomic Energy Agency (IAEA), adopted June 12, 1968, https://www.iaea.org/publications/documents/treaties/npt.

10   "Détente and Arms Control, 1969–1979," Department of State, n.d., https://history.state.gov/milestones/1969-1976/detente.

11   Lawrence S. Wittner, "The Nuclear Freeze and Its Impact," Arms Control Today, December 2010, https://www.armscontrol.org/act/2010_12/LookingBack.

12   "NATO and the INF Treaty," NATO, August 2, 2019, https://www.nato.int/cps/en/natohq/topics_166100.htm.

13   "The Comprehensive Nuclear-Test-Ban Treaty," CTBTO Preparatory Commission, signed September 24, 1996, https://www.ctbto.org/our-mission/the-treaty. Even though the CTBT is not yet in force, it is still influential and has helped seriously restrict nuclear testing.

14   George P. Shultz, William J. Perry, Henry A. Kissinger, and Sum Nunn, "A World Free of Nuclear Weapons," *Wall Street Journal*, January 4, 2007, https://www.wsj.com/articles/SB116787515251566636.

became known as the "Four Horsemen" for nuclear disarmament.[15] The article concludes with their endorsement of "a world free of nuclear weapons and working energetically on the actions required to achieve that goal."

In the wake of the Four Horsemen's influential newspaper column, the "Global Zero" international campaign was formed.[16] In Paris in December 2008, a group of over 100 signatories, including international leaders from political, civil, and military spheres, established the Global Zero organization.[17] The goal of Global Zero was and remains exactly that: global (nuclear) zero.

The Global Zero organization urged then-presidents Barack Obama and Dmitry Medvedev for leadership from the United States and Russia toward the elimination of nuclear weapons. Shortly after, on April 1, 2009, Presidents Obama and Medvedev met in London and issued a historic joint statement, committing their two countries to "achieving a nuclear free world."[18] Just a few days later, on April 5, 2009, in his now-famous speech in Prague, Obama declared "clearly and with conviction America's commitment to seek the peace and security of a world without nuclear weapons," and further, that the United States "will seek to include all nuclear weapons states in this endeavor."[19]

Toward someday achieving global nuclear zero, the Obama administration took some policy actions, including "a [New] START agreement [for nuclear arms reductions] with Russia, a revised nuclear posture deemphasizing nuclear weapons, and a range of nonproliferation initiatives."[20] Obama was awarded the 2009 Nobel Peace Prize in acknowledgement for his efforts toward nuclear disarmament.[21] Yet, it was also the Obama administration that began the United States' pivotal modernization program to update its nuclear forces, estimated to cost $1 trillion over the course of three decades.[22]

The preamble of the NPT recites:

> Desiring to further the easing of international tension and the strengthening of trust between States in order to facilitate the cessation of the manufacture of nuclear weapons, the liquidation of all their existing stockpiles, and the elimination from national arsenals of nuclear weapons and the means of their delivery pursuant to a Treaty on general and complete disarmament under strict and effective international control.[23]

---

15   Eben Harrell, "The Four Horsemen of the Nuclear Apocalypse," *Time Magazine*, March 10, 2011, https://science.time.com/2011/03/10/the-four-horsemen-of-the-nuclear-apocolypse/; and David Cortright, "The Four Horsemen against the Bomb," Notre Dame Magazine, December 14, 2001, https://magazine.nd.edu/stories/the-four-horsemen-against-the-bomb/.

16   "About Us," Global Zero, n.d., https://www.globalzero.org/about-us/.

17   Michael E. O'Hanion, "Is a World Without Nuclear Weapons Really Possible?," Brookings Institution, May 4, 2010, https://www.brookings.edu/opinions/is-a-world-without-nuclear-weapons-really-possible/.

18   "Joint Statement by President Dmitriy Medvedev of the Russian Federation and President Barack Obama of the United States of America," The White House (Obama), April 1, 2009, https://obamawhitehouse.archives.gov/the-press-office/joint-statement-president-dmitriy-medvedev-russian-federation-and-president-barack-.

19   "Remarks By President Barack Obama In Prague As Delivered," The White House (Obama), April 5, 2009, https://obamawhitehouse.archives.gov/the-press-office/remarks-president-barack-obama-prague-delivered.

20   Francis J. Gavin, "8. Global Zero, History, and the 'Nuclear Revolution'," in Francis J. Gavin, *Nuclear Statecraft* (Ithica, NY: Cornell University Press, 2014), 232, doi:10.7591/9780801465765-010.

21   Steven Erlanger and Sheryl Gay Stolberg, "Surprise Nobel for Obama Stirs Praise and Doubts," *New York Times*, October 9, 2009, https://www.nytimes.com/2009/10/10/world/10nobel.html?ref=global-home.

22   Philip Ewing, "Obama's Nuclear Paradox: Pushing For Cuts, Agreeing To Upgrades," NPR, May 25, 2016, https://www.npr.org/sections/parallels/2016/05/25/479498018/obamas-nuclear-paradox-pushing-for-cuts-agreeing-to-upgrades.

23   "Treaty on the Non-Proliferation of Nuclear Weapons (NPT)," signed July 1, 1968, United Nations, https://www.un.org/disarmament/wmd/nuclear/npt/text/.

The NPT entered into force in 1970, now over 50 years ago, and the NWSs have repeatedly rejected calls to negotiate a timebound treaty for the elimination of nuclear weapons. Some NPT proponents favoring nuclear zero have become increasingly frustrated at the lack of progress on nuclear disarmament by NWSs. Largely a result of this frustration, on January 22, 2021, the Treaty on the Prohibition of Nuclear Weapons (TPNW) entered into force.[24] The TPNW is a "comprehensive set of prohibitions on participating in any nuclear weapon activities," leading toward total abolition.[25] This new international agreement was signed and ratified (currently at 86 signatories and 66 parties) by non-nuclear weapon states (NNWSs) exclusively, notably excluding all NWSs, NATO members, and other countries under other nuclear umbrellas (e.g., Japan and South Korea).

The NWSs and NNWSs have different perspectives on whether nuclear zero is ready to be seriously pursued at the current time.

THE INTERNATIONAL ATOMIC ENERGY AGENCY

The IAEA is the world's preeminent international organization for the peaceful use of nuclear energy.[26] The agency has three broad missions: promoting nuclear science and technology for peaceful purposes, developing standards for nuclear safety to protect humans and the environment, and verifying that nuclear materials and facilities are not being used for non-peaceful purposes.[27] It is this third mission that defines nuclear safeguards (i.e., the safeguarding of nuclear energy from military applications).

The objective of IAEA safeguards is "to deter the spread of nuclear weapons by the early detection of the misuse of nuclear material or technology."[28] Nuclear safeguards are a set of technical measures to verify that nuclear material is not diverted toward military purposes. Through its inspection system, the agency aims to verify that its member states are indeed using nuclear facilities only for peaceful purposes to comply with their legal obligations.

The IAEA is the only organization that monitors the world's nuclear activities, with the exception of intelligence, surveillance, and reconnaissance activities that individual states might perform, which are tied to national technical means and are usually classified state secrets.[29] That is to say: the IAEA is the only international, impartial, intergovernmental, and independent organization that openly and legally monitors nuclear facilities and materials on a global scale. This makes IAEA safeguards a uniquely essential component of the international security system.

For all these reasons, as well as to constrain scope, this paper assumes that the IAEA is *the* organization that—in the hypothetical future of nuclear zero—would be credibly tasked with continuously verifying the global elimination of nuclear weapons.

The agency is held in great esteem worldwide for the unique safeguards mission that it successfully carries out. However, even under the less restrictive non-zero paradigm of today, the IAEA already faces challenges and limitations in conducting its safeguards mission. When IAEA safeguards are

---

24   "Treaty on the Non-Proliferation of Nuclear Weapons (NPT)," signed July 1, 1968, United Nations, https://www.un.org/disarmament/wmd/nuclear/npt/text/.

25   United Nations Office for Disarmament Affairs, "Treaty on the prohibition of nuclear weapons," United Nations, signed September 22, 2017, https://www.un.org/disarmament/wmd/nuclear/tpnw/.

26   "About Us," IAEA, n.d., https://www.iaea.org/about.

27   "The IAEA Mission Statement," IAEA, n.d., https://www.iaea.org/about/mission.

28   "Basic of IAEA Safeguards," IAEA, n.d., https://www.iaea.org/topics/basics-of-iaea-safeguards.

29   That is, these national technical means are usually reserved as classified methods, sparing for special arrangements. One such exception are those interagency inspection teams that verify the nuclear arms control agreements under New START, which are *jointly* performed by the United States and Russia.

limited, nuclear verification is limited. Understanding the agency's current constraints should help inform and motivate verification improvements for nuclear zero.

CURRENT CONSTRAINTS LIMITING NUCLEAR VERIFICATION

MATERIAL DETECTION CONSTRAINTS

By definition, to safeguard nuclear material is to ensure that it is all tracked and accounted for. Considering the real-world limits of detection, this is a daunting task.

A "significant quantity" (SQ) is the approximate threshold quantity of nuclear material where the manufacturing of a nuclear explosive device cannot be excluded. For direct-use special nuclear material, this amounts to 8 kg of plutonium (<80 percent Pu-238) or 25 kg of uranium (>20 percent U-235 enrichment).[30] Ideally, the goal of IAEA safeguards is to detect the potential diversion of a single SQ. Practically, this is not always achievable.[31]

Uncertainties in the total amount of nuclear material measured or sampled are typically on the order of about 1 percent.[32] For a large bulk handling facility, 1 percent of the material inventory or throughput may be larger than one SQ, and "in some cases considerably larger."[33] Today, most of the facilities in this category—those that handle direct-use material in bulk—are owned and operated by declared NWSs and as a result are generally not subject to IAEA safeguards.

Further, the IAEA is aware of the theoretical diversion strategy dubbed "partitioning," whereby fractional quantities of material (less than one SQ) are diverted at different times or from multiple facilities, conglomerating to one SQ over time. This strategy is unattractive and high risk due to the coordination involved between multiple parties. Also, the agency would have multiple opportunities to catch this partitioning scheme because, by its nature, multiple diversions are required. However, if the IAEA attempted to more explicitly and rigorously counter these hypothetical scenarios, it believes that "a politically unacceptable inspection regime would be required," or one more invasive than what has been the norm thus far.[34]

The IAEA acknowledges, fairly, that it would be unreasonable to set final inspection goals that cannot be politically or technically achieved. A report from the agency concludes with the following: "The situation is obviously not entirely satisfactory and the limitations of nuclear materials accounting must be offset by the development of effective containment and surveillance measures and by improving the techniques for measuring and accounting for nuclear materials."[35]

---

30   IAEA, *IAEA Safeguards: Aims, Limitations, Achievements* (Vienna: 1983), https://www-pub.iaea.org/MTCD/publications/PDF/IAEA_SG_INF_4_web.pdf.

31   To make matters worse, some experts have argued that the IAEA's current SQ mass thresholds are inaccurate (i.e., too lenient or too large), limiting their usefulness. It is believed that a nuclear explosive could still be produced even with an amount of fissile material smaller than one SQ (at least as currently defined by the IAEA). Some have advocated that the IAEA SQ values should be *lowered by a factor of eight*, down to only 1 kg of plutonium and 3 kg of enriched uranium. Thomas B. and Christopher E. Paine, "The Amount of Plutonium and Highly Enriched Uranium Needed for Pure Fission Nuclear Weapons," Natural Resources Defense Council, April 13, 1995, https://www.ambienteparco.it/pdf/fissionweapons.pdf. See additional reading and discussions: Braden Goddard, Alexander Solodov, and Vitaly Fedchenko, "IAEA 'significant quantity' values: time for a closer look?," *Nonproliferation Review* 23, no. 5–6 (August 2017), doi:10.1080/10736700.2017.1339934; and Jeffrey Lewis, "Significant Quantities Rant," Arms Control Wonk, March 1, 2012, https://www.armscontrolwonk.com/archive/205028/significantly-wrong-about-significant-quantities/.

32   Note that this is the order-of-magnitude uncertainty for measuring *loose* nuclear material, taken as a notional value for the sake of example. For spent fuel, where the vast majority of direct-use metallic form uranium and plutonium exists, measurement uncertainty is usually lower.

33   IAEA, IAEA Safeguards.

34   Ibid.

35   Ibid.

Timely detection is the second obvious requirement of IAEA safeguards, yet there are also practical difficulties in this area.

The IAEA's Standing Advisory Group on Safeguards Implementation (SAGSI) estimates that Pu-239 or high-enriched uranium, in metallic form, can be converted to the components of a nuclear explosive device on the order of 7 to 10 days.[36] (These "conversion times" are months to years for lower-enriched material in non-metallic forms.) The agency provisionally sets its goals for detection times to be on par with the corresponding conversion times.

However, there are real-world challenges in this area as well. If the IAEA sets its detection time to be the same as the 7-to-10-day conversion time for weapons-grade metallic materials, and if timeliness could only be ensured by taking a complete inventory, then physical inventories would need to be taken perhaps once per week. This is a non-starter. A proper physical inventory requires shutting down a facility's operations (halting the processing of nuclear materials), cleaning equipment, and converting materials into a form for accurate measurement—all of which would take more than a week, meaning that a total loss of production would ensue. Furthermore, operators of nuclear facilities, for economic reasons, are reluctant to shut down more than once or twice per year.

As a result, the IAEA's "timeliness" of detection, as achieved via a full inventory, typically ranges from three months to one year. (At a minimum, the IAEA requires one full physical inventory inspection each year. Some facilities agree to three or four.) Because this is far short of the 7-to-10-day conversion time for facilities housing metallic-form enriched nuclear materials (i.e., spent fuel reprocessing or fabrication plants), other approaches are taken to augment for timely detection. This might include camera surveillance, tamper-proof seals on stored nuclear material, and frequent (every 2 to 3 weeks) partial inventories that are designed to minimize the disruption to plant operations. At some larger plants, the IAEA also requires a continuous inspector presence in order to verify the standard flow of nuclear material and observe any potential abnormalities.[37]

The IAEA believes that the number of countries technically capable of abrupt nuclear materials diversion will continue to grow. New facilities for reprocessing and enrichment are being built worldwide, with more countries joining the nuclear power arena. The agency summarized this timeliness factor in one report: "The IAEA must therefore strive to enhance the effectiveness of safeguards and expand its verification activities at reprocessing and enrichment facilities and at stores of plutonium and highly enriched uranium. Continuous human or instrumental inspection may be necessary as well as improved containment and other surveillance measures."[38]

THE IAEA'S LIMITED STAFFING AND FINANCES

Considering the extent and global scale of its tasking, the IAEA safeguards department is rather small in terms of both personnel and budget.

The IAEA Secretariat, which is the international body that represents the agency's entire staff, is comprised of 2,560 personnel from over 100 countries.[39] A subset of this, the IAEA Department of

---

36   Ibid.
37   Ibid.
38   Ibid.
39   "Employees & Staff: Strength Through Diversity," IAEA, n.d., https://www.iaea.org/about/staff.

Safeguards consists of 875 staff from 95 countries, as of 2020.[40] Of these, about 385 are designated inspectors from around 80 countries, and most of the remainder are analysts.[41]

In 2020, IAEA safeguards accounted for 221,432 SQs of nuclear material, spread throughout 1,321 nuclear facilities and related external locations.[42] The inspectors conducted 2,856 in-field verifications, spending 12,767 combined days out in the field and 2,362 days under Covid-19 quarantine, among numerous other measurements, collections, and other activities. These statistics for global nuclear verification depict its lofty scope, while the IAEA safeguards staffing numbers are arguably humble in comparison.

This discussion naturally translates to the IAEA's finances. For 2020 and 2021, the total annual IAEA budget was €592 million ($592 million).[43] Of this, nuclear verification (one of the IAEA's six major programs) was allotted €149 million ($149 million) annually. Despite growth in the agency's global nuclear scope, the operational regular budget has hovered near this amount for the past decade or so.

In 2013, at the International Safeguards Symposium in Vienna, the then-deputy director general of the IAEA, Yukiya Amano, discussed this budget in his opening statement: "Let me put the issue of funding in perspective by repeating the fact that the annual budget for our safeguards work worldwide—approximately €150 million—is less than the annual budget of the police department of the city of Vienna."[44]

In April 2019, Yukiya Amano (director general of the IAEA at the time) delivered an extensively detailed statement on today's challenges in nuclear verification at the Center for Strategic and International Studies (CSIS) in Washington, D.C. In his remarks, Amano made clear how constraints on staffing and finances limit the international nuclear verification work that the IAEA can conduct:

> The steady increase in the amount of nuclear material and the number of nuclear facilities under IAEA safeguards, and continuing pressure on our regular budget, are among the key challenges facing the Agency today.

Pressure on the regular budget is a particularly serious problem for the IAEA. If our regular budget continues to suffer cuts in the coming years, a reduction in the number of IAEA inspectors will be unavoidable. This could seriously undermine our nuclear verification activities.

VOLUNTARY NATURE OF SAFEGUARDS AGREEMENTS

Perhaps the greatest limitation that the IAEA faces is that much of its safeguards efforts depend on (1) voluntary state acceptance of the minimum inspection regime and (2) voluntarily state agreement to the Additional Protocol.

Those states that are not IAEA member states or not voluntary parties to the NPT—although few in number—do not have any reporting obligation to the IAEA. North Korea, as an example, withdrew from IAEA membership in 1994 after it was found to be in non-compliance with its safeguards

40 "IAEA Safeguards in 2020," IAEA, 2020, https://www.iaea.org/sites/default/files/21/06/sg-implementation-2020.pdf.
41 Sasha Henriques, A Day in the Life of a Safeguards Inspector," IAEA Office of Public Information and Communication, July 27, 2016, https://www.iaea.org/newscenter/news/a-day-in-the-life-of-a-safeguards-inspector.
42 "IAEA Safeguards in 2020," IAEA, 2020, https://www.iaea.org/sites/default/files/21/06/sg-implementation-2020.pdf.
43 IAEA, *The Agency's Programme and Budget 2020-2021* (Vienna: 2019), https://www.iaea.org/sites/default/files/gc/gc63-2.pdf. The IAEA's functional currency is the euro, and it assumes a budget exchange rate of €1.00 to $1.00.
44 IAEA, *Preparing for Future Verification Challenges* (Vienna: IAEA, 2013), https://www-pub.iaea.org/MTCD/Publications/PDF/Pub1611_web.pdf.

agreement.[45] In 2003, North Korea officially withdrew from the NPT.[46] In 2009, all agency inspectors in North Korea were expelled. Since then, little progress has been made, as IAEA inspectors have not been permitted regular physical access to North Korean lands. The IAEA has no enforcement powers, neither to compel a country to sign a treaty or safeguards agreement, nor to compel a country to honor their commitments.

Most IAEA member states are subject to the baseline safeguards as prescribed by the NPT.[47] All NNWSs under the NPT, along with states party to regional nuclear-weapon-free zone treaties, are required to establish comprehensive safeguards agreements (CSAs) with the IAEA.[48] However, under this minimum CSA baseline alone, the scope of the agency's inspection authority is rather limited.

In the 1990s, it came to light that Iraq and North Korea had each been in a state of NPT non-compliance.[49] It was discovered that Iraq had been pursuing a clandestine nuclear weapons program (though no nuclear bomb had yet been built). North Korea had obfuscated its nuclear operations and refused inspectors access to certain sites, likely hiding undeclared plutonium production and diversion. While baseline safeguards had been successful for verifying *declared* (known) nuclear material and facilities, it had become clear that the CSAs on their own did not equip the IAEA to detect *undeclared* (unknown) nuclear activities.[50]

Attempting to fill this gap, the IAEA Board of Governors approved the Additional Protocol (AP) in 1997. The AP is a supplemental individual agreement between the IAEA and a state. As a follow-on to the baseline CSA, the AP grants the IAEA "expanded rights of access to information and locations" within the state.

With a fuller picture of a state's nuclear programs, plans, materials, and trade, the AP allows the IAEA to more effectively verify the absence of undeclared nuclear material and activities.[51] To name a few benefits, the AP affords the agency with fuller information and access to the following:

- All parts of a state's declared nuclear fuel cycle (from uranium mines to nuclear waste), including research and development activities (even without nuclear material);
- Any and all buildings on a nuclear site on short notice (2-hour or 24-hour notice);
- All import and export of nuclear-related equipment and material; and
- Any other areas beyond declared locations for the collection of environmental samples, on an as-needed basis.

45   "The Withdrawal of the Democratic People's Republic of Korea from the International Atomic Energy Agency," IAEA, June 21, 1994, IAEA-INFCIRC/447, https://www.iaea.org/sites/default/files/infcirc447.pdf.

46   "Fact Sheet on DPRK Nuclear Safeguards," IAEA, n.d., https://www.iaea.org/newscenter/focus/dprk/fact-sheet-on-dprk-nuclear-safeguards.

47   Note that three countries that are *not* party to the NPT—India, Israel, and Pakistan—still have individual, item-specific safeguards agreements in force with the IAEA. Ionut Suseanu, "The NPT and IAEA safeguards," IAEA, *IAEA Bulletin* 62, no. 4 (December 2021), https://www.iaea.org/bulletin/the-npt-and-iaea-safeguards.

48   "More on Safeguards agreements," IAEA, n.d., https://www.iaea.org/topics/safeguards-legal-framework/more-on-safeguards-agreements.

49   It is more technically accurate to state that Iraq and North Korea were found by the IAEA to be in violation of their nuclear safeguards agreements, and that it was the UN Security Council (not the IAEA) who made the political decision that these infractions amounted to NPT non-compliance.

50   "Additional Protocol," IAEA, n.d., https://www.iaea.org/topics/additional-protocol. Note that the Iraq example was a failure which motivated reforms to bolster safeguards, while the North Korea example was arguably a success story, in that the application of new safeguards measures is what revealed North Korea's past diversions of material.

51   "Strengthening measures," IAEA, n.d., https://www.iaea.org/topics/additional-protocol/strengthening-measures.

That the AP is a voluntary agreement imposes a significant limitation on the IAEA. Though encouraged, a member state is not obligated to enter into an AP with the IAEA to remain in compliance with the NPT. Through 2020, the IAEA had 184 states with safeguards agreements in force, 136 of which had AP agreements. At the time of this paper's writing, there remain certain states with significant nuclear activities but without an AP agreement.[52] This list of non-AP countries includes Argentina, Brazil, Egypt, Iran, Israel, Syria, Venezuela, and others—among which are nuclear reactors, nuclear materials, and nuclear material processing capabilities.[53]

The IAEA concludes that a CSA and an AP must be in force for the agency to verify that *all* nuclear material in a state remains in peaceful activities. For the year 2020, the IAEA positively concluded that *known* (declared) nuclear material remained in peaceful activities in 103 states. In comparison, the IAEA positively concluded that *all* (declared and undeclared) nuclear material remained in peaceful activities in only 72 states.[54]

In his April 2019 remarks, Director General Amano emphasized how essential the IAEA's AP agreement is for nuclear verification:

> One of the most important developments in the history of nuclear verification was the approval by our [IAEA] Board [of Governors] in 1997 of a new legal instrument—the Additional Protocol.

> The Additional Protocol significantly increases the agency's ability to verify the peaceful use of all nuclear material in a country. Without it, we cannot draw what we call the "broader conclusion" that all nuclear material in a country has remained in peaceful activities.

> The combination of comprehensive safeguards agreement and AP needs to become universal. I constantly encourage all countries that have not yet done so to conclude and implement additional protocols.[55]

GLOBAL ZERO IS NUCLEAR VERIFICATION'S GREATEST CHALLENGE

In a December 1982 magazine bulletin, a former IAEA deputy director general for safeguards wrote the following:

> Any activity in the real world has its limitations and the verification activities of the IAEA are no exception. First of all, safeguards cannot be imposed on any state—certainly not by the IAEA. We should be aware that the safeguards system is unique in international relations. It is the first time in the history of our restless species that sovereign states have voluntarily agreed to the inspection of sensitive facilities by foreign nationals. It is, therefore, no surprise that even NPT states will accept inspections only under clearly stated legal and technical constraints derived from international consensus and spelled out in safeguards agreements. It is politically naïve to assume that substantial changes in the basic documents affecting the rights of states would be accepted in the foreseeable future.[56]

52   "STATUS LIST: Conclusion of Safeguards Agreements, Additional Protocols and Small Quantities Protocols," IAEA, December 31, 2021, https://www.iaea.org/sites/default/files/20/01/sg-agreements-comprehensive-status.pdf.

53   Mark Hibbs, "The Unspectacular Future of the IAEA Additional Protocol," Carnegie Endowment for International Peace, April 26, 2012, https://carnegieendowment.org/2012/04/26/unspectacular-future-of-iaea-additional-protocol-pub-47964.

54   Of course, this is not to say that the remaining tens of IAEA member states diverted nuclear material for military use. Rather, it means that the agency lacked sufficient information/access to positively and definitively declare its total verification of solely peaceful use.

55   Amano, "Challenges in Nuclear Verification."

56   H. Grümm, "Potential and limitations of international safeguards," IAEA Bulletin Supplement 24-0, December 1982, https://

This reiterates how remarkable it is that *any* international nuclear safeguards exist at all. It is noteworthy that, by and large, the world's nations have agreed to relax their absolute state sovereignties and begin some semblance of cooperation on this exceptional nuclear issue. Nevertheless, if nuclear verification is to advance, the world must regularly test ideas for expanded safeguards that may or may not be "politically naïve," which might well shift with the times.

The former deputy director general for safeguards continued: "In many instances, expectations of what safeguards can do are rather inflated, and confrontation with the limitations of safeguards often leads to the other extreme—disappointment and harsh criticism."[57]

Yet, nuclear zero raises expectations of safeguards to the most extreme. In order to effectively promote and guarantee an environment of total nuclear disarmament, it should be uncontroversial to argue that most limitations currently imposed on IAEA safeguards will no longer be acceptable under nuclear zero. More rigorous IAEA authorities and inspection schemes will be required for the credible assurance that the world remains free of all nuclear weapons. The IAEA's constraints in the prior section need to be addressed and corrected in order to accomplish such a task.

All of this is easier said than done. After Obama expressed support for the pursuit of global zero, he was mocked as naïve and idealistic (emphasis mine): "Achieving a world without nuclear arms would require, at a minimum, that nations conclude that they could protect their vital interests without nuclear arms; that *new and very intrusive verification mechanisms were developed and agreed*; and that an *enforcement mechanism against any cheating state have real teeth*—daunting challenges, to be sure."[58]

There is merit to these criticisms. Indeed, the highest standards of nuclear arms control—global nuclear zero—ought to similarly require the highest standards of nuclear verification.

The challenge of nuclear zero will be ensuring that it remains at zero. Even if all the world's nuclear weapons were somehow eliminated right now, the immediate question that each government would nervously ask itself is "Will someone else develop nuclear weapons again?" If any state abandons zero, there could well be dire global security consequences. It is precisely this question, and with it the pervasive fear of re-proliferation, that would almost surely drive nations to become nuclear possessors anew.

In order to keep nuclear zero, the conditions must be created where this question is moot. The only way to eliminate the fear of re-proliferation is universal high confidence in a truly robust nuclear zero verification regime. Any paths for potential re-proliferation, and with them any room for doubts and security uncertainties, must be removed. States must be able to trust that other states could not obtain a nuclear advantage because any such attempts would be swiftly discovered and thwarted.

Under nuclear zero, nation-state governments must be convinced that a wide-sweeping, comprehensive verification system is in place—a credible and effective monitor for all potential nuclear proliferation activities—to catch (and stop!) would-be proliferators before a nuclear weapon is redeveloped. Toward this end, this paper will now provide several recommendations for consideration. These proposals are a collection of new monitoring technologies, expanded resources,

---

www.iaea.org/sites/default/files/publications/magazines/bulletin/bull24-0/24003484043su.pdf.

57   Ibid.

58   Steven Pifer, "10 years after Obama's nuclear-free vision, the US and Russia head in the opposite direction," Brookings Institution, April 4, 2019, https://www.brookings.edu/blog/order-from-chaos/2019/04/04/10-years-after-obamas-nuclear-free-vision-the-us-and-russia-head-in-the-opposite-direction/.

higher verification standards, and new or upgraded legal procedures, many of which might be paradigm shifts that raise the bar of previously established safeguards. The intent is to dramatically strengthen international verification capabilities in order to rapidly detect (and quickly and decisively respond to) any nuclear proliferant activities.

PROPOSALS FOR RIGOROUSLY STRENGTHENED SAFEGUARDS TO PREPARE FOR NUCLEAR ZERO

Staffing and budget for IAEA safeguards should be massively expanded. This is the foundational proposal that will relate to and enable many of the subsequent proposals for strengthening nuclear verifications. With its current personnel numbers and funding, the IAEA already has very little margin of error in conducting safeguards at the present level of fidelity; to attempt verification of nuclear zero, these resources would be drastically insufficient.

The IAEA's safeguards staffing and budget should both be healthily increased. Perhaps the United Nations must mandate that all nations contribute a certain minimum percentage of their country's gross domestic product, and perhaps the world's current NWSs should face the largest financial burden. It is difficult to propose specific numbers for staffing and budget within the constrained local context of this paper. However, to verify nuclear zero, there will be an essential need for the standing presence of several thousand inspectors, which in turn will require a corresponding (and proportional) increase of analysts for support. This will be discussed further below.

Regarding the IAEA safeguards budget to support these efforts, significantly more analyses and details are needed. For reference, note that the world's nine nuclear-armed nations spend a combined $73–100 billion annually on nuclear weapons.[59] This entirely dwarfs the IAEA's current yearly budget for safeguards—$162 million. This would suggest that, in a world without nuclear weapons, there should be ample room to accommodate for fair and sizeable increases in the IAEA budget and staffing.

Basic safeguards (NPT+CSA) and the AP must both become mandatory. Both membership in the NPT/IAEA and the AP agreement must be made mandatory for all nations to maintain nuclear zero. The international community needs North Korea to rejoin the NPT and provide the IAEA access to its nuclear facilities and materials. India, Israel, and Pakistan are also noteworthy in that they have never been signatories of the NPT; this must be resolved. Further, the international community must work with the remaining holdout states that currently lack an AP agreement with the IAEA—perhaps especially Argentina, Brazil, Egypt, Iran, Israel, Syria, Venezuela, and other nations with nuclear facilities.

Without the NPT and without an AP with the IAEA—that is, without the very basics of nuclear safeguards for inspections and verifications—there can only be limited outside speculation of how a country is operating its nuclear program. This is in clear and stark opposition to the standards that nuclear zero requires. A state cannot have the option to not comply or withdraw from any of these legal commitments; there must be mechanisms for the rigorous enforcement of membership and

---

59   The bounds of this estimated range in global nuclear weapons spending come from two separate sources: Alicia Sanders-Zakre, *Enough is Enough: 2019 Global Nuclear Weapons Spending* (Geneva: International Campaign to Abolish Nuclear Weapons (ICAN), May 2020), https://www.icanw.org/report_73_billion_nuclear_weapons_spending_2020; and Bruce G. Blair and Matthew A. Brown, "World Spending on Nuclear Weapons Surpasses $1 Trillion per Decade," Global Zero, June 2011, https://www.globalzero.org/wp-content/uploads/2020/01/GZ-Weapons-Cost-Global-Study.pdf. Two important caveats: (1) these estimates are inferred from open-source data in the public domain, which might be inaccurate or not the entire picture (i.e., classified budgets), and (2) these estimates were calculated by ICAN and Global Zero, two organizations that actively promote total nuclear disarmament.

compliance on all matters of safeguards, for all countries. Discussed in more detail below, for nuclear zero to hold, a country's failure to fully cooperate with the IAEA would have to be interpreted by the international community as not merely a diplomatic faux pas, but rather more harshly as akin to an act of war with a strong call to action.

No-notice inspections, continuous monitoring, and continuous IAEA presence should become standard. The final stage of the Global Zero campaign's vision for nuclear zero, "Phase 4," concludes (emphasis mine): "Building on decades of successful monitoring and verification programs, implementation [of zero] is carefully overseen by international institutions and *confirmed by constant surveillance and on-site, no-notice inspections—indefinitely*."[60]

For nuclear zero to be maintained, the world's nuclear facilities must be continuously monitored. Correspondingly, the IAEA's nuclear verification activities must increase in frequency, duration, and presence. If surveillance is only periodic, there could be a chance (however small) of nuclear material being diverted for military use. The IAEA could make use of several verification instruments (technologies and strategies) to accomplish continuous monitoring.

First, the AP currently allows for short-notice inspections (2 hours or 24 hours) of nuclear facilities by the IAEA. States must agree to an updated provision enabling zero-notice inspections. Even with the knowledge that nuclear facilities are complex and sluggish—generally unable to alter their running conditions on the order of hours without leaving traces of their operation history in the materials, environment, and logbooks that inspectors can detect—it would be better for the IAEA to conduct no-notice inspections. This change could benefit nuclear zero in multiple ways. It could serve as a psychological deterrent, dissuading would-be plant operators from engaging in unorthodox activities out of fear of being caught in the very act. No-notice inspections could also prevent attempts to cover up or hide proliferant acts.

Second, multiple devices that are constantly recording could be installed in all nuclear facilities. This objective data can augment the eyes of IAEA inspectors. Cameras are not unprecedented; indeed, some nuclear plants today already have IAEA-installed cameras, such as Iran's centrifuge manufacturing facility. In 2019, IAEA director general Amano stated: "We [the IAEA] increasingly monitor nuclear facilities remotely in real time, using permanently installed cameras and other instruments. The agency collects and analyzes hundreds of thousands of images captured daily by our surveillance cameras installed in numerous nuclear facilities."[61]

Radiation detectors are another recording device that could be IAEA owned and operated and installed at all nuclear facilities. The radiation signatures of fissile isotopes (e.g., U-235, Pu-239, U-233) are of particular interest, and detector measurements could help verify the presence or absence of these and other materials of interest, as well as track the activity of the plant during its operations.

Third, for the best assurance of nuclear zero, at least one IAEA inspector could be permanently stationed at each of the world's nuclear facilities (if possible). At the time of this paper's writing, according to IAEA global databases, there are:

- 696 nuclear power reactors (439 currently operational) in 32+ countries;[62]

---

60 "Reaching Zero," Global Zero, n.d., https://www.globalzero.org/reaching-zero/.

61 Amano, "Challenges in Nuclear Verification."

62 "In-Operation and Long-Term Shutdown," IAEA Power Reactor Information System, July 22, 2022, https://pris.iaea.org/PRIS/WorldStatistics/WorldStatisticsLandingPage.aspx.

- 842 nuclear research reactors (220 currently operational) in 70 countries;[63] and

- 779+ nuclear fuel cycle facilities involved in materials processing (e.g., uranium ore processing, conversion, enrichment, fuel fabrication, and reprocessing) in 55+ countries.[64]

These numbers might be considered lower bounds for the future, as interest in nuclear energy continues to grow. As mentioned above, several thousand IAEA inspectors (and supporting analysts) would be needed for permanent stationing at as many sites as possible. At a minimum, a continuous inspector presence should be mandatory at sites that deal with direct-use uranium and plutonium materials, such as nuclear reactors and facilities for uranium enrichment and fuel reprocessing.

Note that this would be in addition to, not a replacement of, the standard visiting IAEA inspectors that perform full inventory inspections once or a few times per year per site. There is precedence for a continuous inspector presence at certain sites: "At some larger plants the IAEA also requires the continuous presence of inspectors to verify the internal flow of nuclear material, and thus to achieve the timeliness goal [for the detection of potential undeclared diversion of material]."[65]

Again, quoting Amano's 2019 remarks, "technology can never be a full substitute for the presence of experienced inspectors on the ground."[66] An IAEA inspector constantly on-site, at the major nuclear sites (or all sites if possible), ought to be a significant boon toward building international confidence in verifying nuclear zero. It would probably be wise for these on-site inspectors to be rotated in their stations on a periodic basis, perhaps every few months or annually. This rotation would remove the (admittedly low) chance of bribery or unethical and corrupt conduct, which might tempt a nation into thinking that it could get away with non-compliance. Again, every positive and proactive measure must be taken for the credibility of verifying the lack of nuclear weapons.

This would probably even necessitate an IAEA inspector be stationed on each nuclear-powered military vessel, namely submarines and aircraft carriers.[67] Understandably, governments will resist the idea of a foreign national presence on board military vessels. Care must be taken that IAEA inspectors maintain strict nondisclosure agreements to ease concerns by ensuring the protection of confidential information.

As an additional benefit, this continuous human presence could help preemptively address reasonable concerns that greatly expanded IAEA inspectors and inspections would be economically crippling for nuclear facilities. It would be a non-starter for countries to shut down their nuclear material processing at facilities several times per year, for weeks at a time, for full physical inventories by inspectors. However, an IAEA inspector continuously on-site to monitor for steady plant operations and material flow should allow plant shutdowns for full physical inventories to remain relatively infrequent and not become economically debilitating.

---

63 "Research Reactor Database," IAEA, n.d., https://nucleus.iaea.org/rrdb/#/home.

64 "Number of Nuclear Fuel Cycle Facilities," IAEA INFCIS, n.d., https://infcis.iaea.org/NFCIS/Facilities.cshtml.

65 IAEA, IAEA Safeguards.

66 Amano, "Challenges in Nuclear Verification."

67 As a related aside: AUKUS—the trilateral security pact between Australia, the United Kingdom, and the United States that seeks to provide Australia with nuclear-powered submarines—is set to be an interesting challenge for nuclear safeguards, and the IAEA has formed a special team to investigate the legal implications. Julian Border, "IAEA chief: Aukus could set precedent for pursuit of nuclear submarines," *The Guardian*, October 19, 2021, https://www.theguardian.com/world/2021/oct/19/iaea-aukus-deal-nuclear-submarines.

Establish unparalleled capabilities to detect undeclared sites. No-notice IAEA inspections, continuous monitoring via on-site cameras and radiation detectors, and a constant IAEA inspector presence on-site all seem to be excellent strategies for declared nuclear facilities and materials, but what about for undeclared sites? To best oversee global zero, the IAEA should be afforded a variety of new and improved tools and capabilities—some of which will be sizeable asks—to survey the world for potential covert nuclear sites.

First, the IAEA should have abundant access to near-real-time global earth imagery data at detailed resolutions. In 2020, the IAEA acquired 1,264 commercial satellite images. For nuclear zero, the IAEA will likely require significantly more earth imagery data, which is one of the many reasons why the agency would make use of a robustly improved staffing and budget. If sufficient in quality and quantity, this imagery could continue to come from commercial satellites; otherwise, it might be worth providing the IAEA with its own network of satellites. With this data, the IAEA could monitor for unusual or suspicious activities, both at known nuclear sites and any potentially undeclared sites. For example, both large gaseous-diffusion uranium-enrichment plants and plutonium-producing nuclear reactors typically have large optical-spatial or thermal-infrared footprints that can be detected remotely via satellite imagery.[68] However, nuclear enrichment facilities can be disguised in their appearance or concealed underground, and slower or more expensive enrichment technologies exist that broadcast a minimal environmental presence, which means that aerial information alone is insufficient.

Second, IAEA inspectors should be unrestricted in their access to a host country's lands. Unrestricted environmental sampling near known nuclear sites should be a given. Unrestricted access to buildings on declared nuclear sites should also be a given. Further, environmental sampling of various other parts of the country should be performed on an as-needed basis; this should not be prevented. Further still, countries should permit IAEA inspectors access even to sites and facilities that are not known to be nuclear-related. In essence, this means that the agency inspectors would be able to examine any potential area of a country or building in question if they have reason to believe that hidden nuclear activity is being conducted there. This is sure to be a very hard sell. Without permitting the IAEA inspectors to access all the areas that it requests, the agency might not be able to fully investigate anomalous environmental samples or third-party tips and other information. If certain areas are "off limits," it becomes troublesome to positively vet nuclear zero.

Third, the employments and various activities of former nuclear weapons scientists might be tracked. This is probably a task for the governments of the former NWSs or the United Nations rather than a task for the IAEA, but it might be something that needs to be reported to the IAEA for awareness. Technical expertise does not vanish and is not unlearned. It might be worth verifying that this select group of individuals is not disseminating nuclear weapons knowledge.

Enhance economic and military enforcement of nuclear safeguards. Under the current safeguards paradigm, safeguards are somewhat loosely enforced. The IAEA does not have enforcement powers, perhaps rightly so, as it is not and should not be an enforcement body. The IAEA should remain a neutral and independent organization and should not be asked to enforce their own rulings. Rather, the United Nations (or something akin to a world government) should be (seriously) empowered

---

68   Hui Zhang and Frank N. von Hippel, "Using Commercial Imaging Satellites to Detect the Operation of Pluto-nium-Production Reactors and Gaseous-Diffusion Plants," *Science & Global Security* 8, no. 3 (2000): 261–313, doi:10.1080/08929880008426479.

as the authority to legislate, judicially process, and, most importantly, enforce matters pertaining to nuclear proliferation, based upon IAEA evidence.

Arguably, the lack of an uncompromising enforcement mechanism behind the IAEA has enabled certain countries—North Korea, Iran, and Syria, to name a few—to not cooperate with the agency. IAEA inspectors have been restricted from accessing certain facilities, prevented from installing monitoring equipment, or outright barred from entering borders due to nationality prejudices, for example. Additionally, the inspectors' host countries do not always proffer substantial information on their facility operations. These are but some examples of how nations can and have impeded the IAEA's nuclear verification mission. As Director General Amano put it in 2019: "Such cases serve as a reminder of the extent to which safeguards implementation depends on cooperation by the countries concerned. They also illustrate the need for states to engage in serious negotiations in cases of non-compliance in order to make it possible for the agency to carry out its verification work."[69]

For nuclear zero, such actions and other behavior cannot be permitted to negatively impede the IAEA's work; nuclear safeguards must be rigidly enforced through uncompromising diplomacy, economic sanctions, and, if necessary, military action. If a country refuses to cooperate with reasonable IAEA requests, or generally impedes the agency's standard operations, there needs to be serious consequences. An atmosphere of intolerance must be established in the international community—intolerance of any nation restricting the IAEA safeguards personnel from completing their work.

With necessary international consensus, if a nation interfered with the IAEA on a relatively minor level, it could be quickly met with international condemnations and harsh sanctions to quickly reverse this behavior. If the circumstances are more severe (e.g., if the IAEA finds a country to be diverting nuclear materials for military purposes, if a research and development program for nuclear weapons is discovered, or if there are strong indications of an undeclared nuclear site that the IAEA is prevented from inspecting), then there must be an even more severe, overwhelming international response. Should diplomacy and extreme economic sanctions fail or seem unlikely to succeed promptly, an expedient international military effort would have to be considered. While military action might be extreme, what other alternative is available? The consequences of nuclear zero failing are *more* extreme and could well be catastrophic.

Nuclear zero requires that action be taken quickly. For a country with uranium enrichment or fuel reprocessing sites with material reserves, the breakout time (i.e., the time needed to procure enough weapons-grade uranium or plutonium for a bomb, an SQ in the IAEA's nomenclature) can be incredibly short. In a recent example in the worst-case scenario in September 2021, Iran's breakout time was estimated to be as short as a month.[70] A world without nuclear weapons cannot afford to accommodate delays in the IAEA's continuous verification work. Upon the acceptance of vetted evidence from the IAEA, proliferation countermeasures from the United Nations should activate automatically in order for enforcement to move at a sufficiently brisk pace.

International consensus is essential to quickly and strictly enforcing IAEA safeguards for nuclear zero. For this to be possible, all nations must be on even ground. Enforcement of nuclear zero should supersede and disregard other alliances or treaties. There should be no exceptions, no exemptions, and no routes either to circumnavigate non-proliferation obligations or to shirk responsibility to

---

69   Amano, "Challenges in Nuclear Verification."
70   "IAEA Report: Breakout Time Shortens," United States Institute of Peace, *Iran Primer*, September 15, 2021, https://iranprimer.usip.org/blog/2021/sep/15/iaea-report-breakout-time-shortens.

respond to potential safeguards violations. This also means that the special privileges otherwise afforded to the permanent members of the UN Security Council cannot apply in these areas. The UN Security Council should not have veto powers on matters of safeguards for nuclear verification in a global-zero environment. Interestingly, there is historical precedence to this idea. In 1946, the United States presented its Baruch Plan to the UN Atomic Energy Commission, which proposed the decommissioning of all U.S. nuclear weapons, alongside establishing international control of nuclear technology. In his plan, Baruch explicitly states the following:

> The [United Nations] Charter permits penalization only by concurrence of each of the five great powers [permanently seated on the Security Council]—the Union of Soviet Socialist Republics, the United Kingdom, China, France, and the United States.
>
> I want to make very plain that I am concerned here with the veto power only as it affects this particular problem [of violating nuclear disarmament]. There must be no veto to protect those who violate their solemn agreements not to develop or use atomic energy for destructive purposes.
>
> The bomb does not wait upon debate. To delay may be to die. The time between violation and preventive action or punishment would be all too short for extended discussion as to the course to be followed.[71]

Acknowledge the deficiencies of these proposals. With regards to these proposals, there are several obvious limitations to immediately acknowledge.

First, this listing of proposals is not comprehensive, nor is this listing of deficiencies.

Second, these proposals generally take for granted a hypothetical world without nuclear weapons as a starting point. The focus was on how to ensure that a given nuclear-zero world is maintained. In generating these verification-related ideas, not much attention was given to the other question of how to reach global zero in the first place. There are probably other ideas and approaches that will be needed for this phase. In other words, on the way down to zero, how can the world's current nuclear powers cooperate and mutually verify that each nation is truly dismantling its warheads?[72]

---

71   Bernard Baruch, "The Baruch Plan," AtomicArchive.com. June 14, 1946, https://www.atomicarchive.com/resources/documents/deterrence/baruch-plan.html.

72   Research is ongoing to develop methods for nuclear weapon treaty verification. Direct warhead dismantlement verification is an interesting and challenging problem: how can the presence or absence of a real, authentic nuclear warhead be confirmed without revealing the classified secrets of the warhead's design? The Laboratory for Nuclear Security and Policy at MIT is one of the research groups investigating this problem: "Treaty Verification with Resonant Phenomena," Laboratory for Nuclear Security and Policy, n.d., http://lnsp.mit.edu/treaty-verification/. To ever reach nuclear zero, the IAEA's role in nuclear disarmament verification must be more firmly defined and bolstered. One obvious roadblock to zero that could be addressed in the near term is that NWS nuclear weapons facilities are not monitored currently. (This is largely because these facilities are de facto already known to be producing weapons materials.) As a good-faith effort toward eventual disarmament, nuclear-armed states could subject their excess military fissile material to safeguards. This would be a significant confidence-building measure, promoting greater transparency while also improving the IAEA's access to sites and materials. In fact, there is recent historic precedence for these sort of NWS safeguards. The so-called "Trilateral Initiative" between the Russian Federation, the United States, and the IAEA placed excess weapons material under IAEA supervision. This program ran from 1996 to 2002, and it investigated verification schemes for the IAEA to keep an accountancy of nuclear material in classified warheads or weapons components. A Model Verification Agreement was even negotiated between the parties; this is a draft legal framework for how the IAEA could verify an NWS's release of fissile material from defense programs (i.e., the verification that special nuclear material was irreversibly removed from nuclear warheads), perhaps in a scenario where an NWS is dismantling some or all of its weapons. This is precisely the sort of "good-faith" negotiation that Article VI of the NPT asks the NWS to pursue toward the elimination of nuclear weapons. It is also a strong example of how the IAEA's role could be quite naturally expanded to include the verification of nuclear disarmament, in addition to its traditional nonproliferation verification activities. This and more is discussed in: Laura Rockwood, "The Trilateral Initiative: The Legal and Financial Issues," IAEA, October 17, 2014, https://inis.iaea.org/

That said, the argument might be made that nuclear zero could become a more appealing idea for countries to more sincerely consider if the necessary verification mechanisms are already in place prior to their reductions to zero. Thus, this paper's conglomeration of proposals for maintaining zero might also assist in motivating countries to eliminate their nuclear weapons to join zero.

Third, many of these proposals are indeed "politically naïve," at least for now. The highest level of rigor—that is, a high level of invasiveness and cost—is required to achieve global nuclear zero. International relations are complex, and if these safeguards proposals were simple, they would already have been agreed upon. However, many of these ideas or related, less-stringent approaches could also be useful in the broader arms control sense (i.e., short of nuclear zero). An even more proficient safeguards verification system could make further arms control agreements (of the non-absolutist, non-abolitionist variety) more palatable. Reinforcing and furthering the current IAEA safeguards scheme could naturally beget greater interest in collaborative dialogues and agreements on nuclear weapons reductions between states. Taking this longer view, perhaps on the scale of decades, it has been argued that by the time a nuclear-weapons-free world is finally realized, "there will be far greater experience in sustaining complex verification systems for weapons of mass destruction, as well as in other security-related areas."[73]

Fourth, in spite of all of these powerful technologies and procedures for verifying nuclear activities, with redundancies, they probably should not be regarded as the panacea that would wholly guarantee that no nuclear weapons could ever be produced. It is usually prohibitively expensive and difficult to satisfy an "always" or "never" objective. It seems more likely that there would still remain a possibility of a nation violating nuclear zero by somehow reacquiring nuclear weapons. There might be some unforeseen pathway, circumventing these verification systems, for procuring weapons-grade nuclear material and assembling it into a nuclear explosive package. Or, even if these verification systems work as advertised, they might still be ultimately ineffective. Even if a country is caught diverting fissile material to build a nuclear weapon early in the act, sanctions and even military action might be insufficient or too slow to stop them in time. Further still along these lines, in the hypothetical case of a major war in a nuclear-zero world, there is a salient point that merits further consideration:

> Common sense would suggest that immediately following the outbreak of hostilities—if not in the run-up to the war itself—every previous nuclear power would make a rapid dash to reconstruct their nuclear forces in the shortest amount of time. But as deadly as a modern conventional war would be in a nuclear free world, the real danger is that it wouldn't remain conventional. Along with making great power conflict far more likely, global nuclear disarmament offers no conceivable mechanism to ensure that such a war would remain non-nuclear.[74]

Fifth, this paper's proposals and ideas would benefit from significantly more cross-examination.

collection/NCLCollectionStore/_External/safeguards/symposium/2014/home/eproceedings/sg2014-papers/000275.pdf; Thomas E. Shea, "Report on the Trilateral Initiative: IAEA Verification of Weapon-Origin Material in the Russian Federation & United States," IAEA, *IAEA Bulletin* 43, no. 4 (2001), https://www.iaea.org/sites/default/files/publications/magazines/bulletin/bull43-4/43403054953.pdf; and Vertic: *The IAEA and Nuclear Disarmament Verification* (London: September 2015), https://www.vertic.org/media/assets/Publications/VM11%20WEB.pdf.

73   Suzanna van Moyland, *Sustaining A Verification Regime in a Nuclear Weapon-Free World* (London: VERTIC, June 1999), http://www.vertic.org/media/Archived_Publications/Research_Reports/Research_Report_4_Van_Moyland.pdf.

74   Zachary Keck, "A Global Zero World Would Be MAD," *The Diplomat*, March 24, 2014, https://thediplomat.com/2014/03/a-global-zero-world-would-be-mad/.

CONCLUSION

In discussing hopes for North Korea's denuclearization, Director General Amano remarked:

> It is important that any agreement on denuclearization is accompanied by an effective and sustainable verification mechanism. The IAEA, with its long experience and well-established practices, is the only international organization that can conduct verification and monitoring activities in an impartial, independent and objective manner. This would help to make the implementation of any agreement sustainable.[75]

It would probably be folly to assume that global nuclear zero, a most stringent goal, would naturally persist—that is, not without a bold, well-equipped, and effective international safeguards regime for nuclear verification that is *commensurately stringent*. The sheer extent and intrusiveness of a nuclear zero verification system demands that it be enforced by, essentially, a world government with universal authority; certainly, this is a most difficult ask.

The world's nations ought to seriously invest in further policy instruments for international nuclear verification. Additional technologies and strategies to remove gaps in safeguards would better enable prospects for future multilateral arms control treaties on nuclear weapons, irrespective of whether this path leads all the way down to zero or not.

As the famed Russian proverb goes, and as President Ronald Reagan employed on several occasions in dialogues with the Soviet Union on nuclear disarmament: "Trust, but verify."

75   Amano, "Challenges in Nuclear Verification."

Nuclear Temperature Rising?

The Impact of Climate Change on Nuclear Stability

By John Madeira[1]

INTRODUCTION

In the decades following the first and only use of nuclear weapons at the end of World War II, global leaders realized that the sheer destructive power of the weapons posed an existential threat to life on earth. During a radio address in 1982, U.S. president Ronald Reagan reinforced the point by saying "Those who've governed America throughout the nuclear age and we who govern it today have had to recognize that a nuclear war cannot be won and must never be fought."[2] As the world phased out of a Cold War mentality in the 1990s and 2000s, the threat of nuclear weapons remained present but was much less of a focal point of international security policy. In a post-Cold War world, the United States and international community began to confront a new wave of security challenges in transnational threats, such as violent extremism and, more recently, the borderless threat of climate change. While some may argue against climate change posing an "existential threat" to the world, U.S. president Joe Biden would disagree.[3] During the first week of the Biden administration, the president, speaking on a domestic plan to combat climate change, stated "that's why I'm signing today an executive order to supercharge our administration['s] ambitious plan to confront the existential threat of climate change. And it is an existential threat."[4] While nuclear war and climate change each pose their own unique set of security challenges, the two also interact with each other in various ways. Of note for this paper, climate change is helping to reshape the global geopolitical landscape,

---

1   John Madeira is a project lead at CRDF Global in Arlington, VA, working on Cooperative Threat Reduction Programs. Views presented are his own and do not represent the views of CRDF Global or its partner organizations.

2   Ronald Reagan, "Radio Address to the Nation on Nuclear Weapons," (speech, April 17, 1982), https://www.reaganlibrary. gov/archives/speech/radio-address-nation-nuclear-weapons.

3   Marc A. Thiessen, "Opinion: Climate change is not an 'existential threat,'" *Washington Post*, https://www.washingtonpost. com/opinions/2021/11/02/climate-change-is-not-an-existential-threat/.

4   Joseph R. Biden, "Remarks by President Biden Before Signing Executive Actions on Tackling Climate Change, Creating Jobs, and Restoring Scientific Integrity," (speech, The White House, January 27, 2021), https://www.whitehouse.gov/brief-ing-room/speeches-remarks/2021/01/27/remarks-by-president-biden-before-signing-executive-actions-on-tackling-cli-mate-change-creating-jobs-and-restoring-scientific-integrity/.

including in the Arctic, in a manner that could have implications for nuclear stability between the United States and Russia, especially as Washington refocuses its security and defense priorities on strategic competition for the first time since the collapse of the Soviet Union in 1991.

ARCTIC CLIMATE CHANGE AND STRATEGIC COMPETITION

The effects of climate change are being felt all over the world and in many forms. For example, in the Sahel region of West Africa, climate change is causing desertification of arable land at a rate of "almost a mile per year in the last decade or so."[5] This is increasing competition for scarce resources, such as land, between farming and herding communities and has even resulted in violence in recent years. On the other hand, climate change is affecting parts of South Asia with increased flooding and longer monsoon seasons, which can destroy farmland, cause food shortages, and even be deadly. In the Arctic, climate change is seen most notably in diminishing levels of sea ice as global temperatures rise. According to the National Snow and Ice Data Center, the Arctic has warmed at a rate approximately twice that of the rest of the planet, leading to a phenomenon called Arctic amplification.[6] Diminishing sea ice levels can be seen by looking at the annual average sea ice extent, or the total ocean area that has an ice concentration of 15 percent or more. For example, the data shows the average Arctic sea ice extent in September has dropped 800,000 square miles (just larger than the square mileage of Mexico) between 1979 and 2021.[7]

Figure 1: Average September Arctic Sea Ice Extent

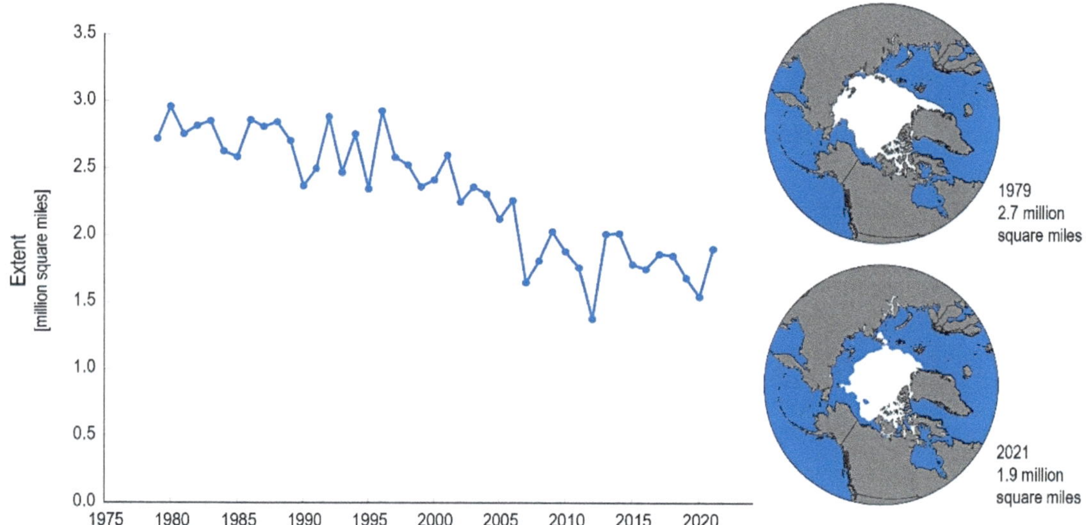

Source: "Arctic Sea Ice Extent," U.S. Global Change Research Program, https://www.globalchange.gov/browse/indicators/arctic-sea-ice-extent#metadata-downloads with original figures found in "Lingering seashore days," *Arctic Sea Ice News & Analysis*, National Snow & Ice Data Center (NSIDC), October 5, 2020, http://nsidc.org/arcticseaicenews/2020/10/ and "Sea Ice Spatial Comparison Tool," *Arctic Sea Ice News & Analysis*, NSIDC, http://nsidc.org/arcticseaicenews/sea-ice-comparison-tool/.

---

5   Caitlin Werrell and Francesco Femia, "AFRICOM Commander on Climate Change: Sahel Receding Almost a Mile Per Year," Center for Climate and Security, March 16, 2018, https://climateandsecurity.org/2018/03/africom-commander-on-climate-change-sahel-receding-almost-a-mile-per-year/.

6   "Climate Change in the Arctic," National Snow & Ice Data Center, https://nsidc.org/cryosphere/arctic-meteorology/climate_change.html.

7   "Arctic Sea Ice Extent," U.S. Global Change Research Program, https://www.globalchange.gov/node/3652.

As climate change melts Arctic sea ice, the waters of the Arctic become navigable for longer stretches of the year, which can have serious economic ramifications by opening new shipping routes, allowing increased access to ports, and improving mineral and resource accessibility. For example, the ability to ship goods via the Northern Sea Route (NSR) could cut shipping time between Europe and Asia by as much as half and provide substantial economic benefits, especially for Russia, which currently uses the NSR to ship more than 30 million tons of hydrocarbons and other resources per year.[8] Additionally, new port development can allow for increased maritime shipping activity and help turn the abundance of natural resources in the Arctic into economic benefits for whichever country gets there first. A 2008 estimate of these resources projected 90 billion barrels of oil, 1.7 quadrillion cubic feet of natural gas, and 44 billion barrels of natural gas liquids.[9]

While the Arctic was once considered a place of cooperation and low tensions, it has since become a geopolitical hotspot of strategic competition. One of the primary ways geopolitical competition plays out in the Arctic is dueling militarization between Washington and Moscow. Arctic militarization as a form of geostrategic competition was prominent during the Cold War when both the United States and Soviet Union looked to secure their northern flanks. Both sides developed strategic bombers and ballistic missiles capable of delivering nuclear weapons over the North Pole. Washington further militarized the Arctic by building radar systems in Alaska (with U.S. ally Canada also building radar systems on Canadian soil), building military bases in Greenland and Iceland, hosting U.S. (and other NATO allied troops) in Norway, using the Arctic as a testing ground for air- and sea-launched cruise missiles, and keeping B-52 strategic bombers armed with thermonuclear bombs on continuous airborne alert. On the Soviet side, Moscow used the Arctic as territory to conduct nuclear tests and house the country's Northern Fleet, which would be the largest Soviet fleet by the early 1960s and home of 60 percent of the Soviet Union's submarine-based strategic forces.[10]

Following the end of the Cold War, militarization of the Arctic fell by the wayside in both Washington and Moscow. Then, in 2007, Russian president Vladimir Putin re-prioritized the Arctic as part of his grand vision of restoring Russia's great power status. Russia's military presence in the Arctic seeks to achieve three main objectives:

1. Enhance homeland defense, especially with regards to deterring a foreign attack in the Arctic as the region begins to attract international attention and investment;

2. Ensure Russia's economic prosperity and secure its economic future; and

3. Create a base of power projection, particularly in the North Atlantic.[11]

8   Gleb Stolayrov, "Russia aims for year-round shipping via northern sea route in 2022 or 2023," Reuters, October 11, 2021, https://www.reuters.com/world/europe/russia-aims-year-round-shipping-via-northern-sea-route-2022-or-2023-2021-10-11/.

9   "Circum-Arctic Resource Appraisal: Estimates of Undiscovered Oil and Gas North of the Arctic Circle," U.S. Geological Service, July 23, 2008, https://pubs.usgs.gov/fs/2008/3049/fs2008-3049.pdf.

10   Kristina Spohr and Daniel S. Hamilton, "From Last Frontier to First Frontier: The Arctic and World Order," in *The Arctic And World Order*, Kristina Spohr and Daniel S. Hamilton, eds. (Washington, DC: Johns Hopkins University, 2020), 3.

11   Matthew Melino and Heather A. Conley, "The Ice Curtain: Russia's Arctic Military Presence," CSIS, March 26, 2020, https://www.csis.org/features/ice-curtain-russias-arctic-military-presence.

Figure 2: Russia's Kola Peninsula

Note: Annotation by author.

Source: "Russia's Kola Peninsula," CIA, The World Factbook, n.d., https://www.cia.gov/the-world-factbook/countries/russia/map.

As such, the Russian Federation has dedicated substantial resources toward remilitarizing the region. Russia's Northern Fleet, based in the Kola Peninsula bordering Finland, is now home to 70 percent of its submarine-based strategic deterrent, 39 attack submarines that carry dual-use Kalibr cruise missiles, and approximately 40 surface ships, including destroyers and cruisers.[12] Additionally, Russia has opened more than 50 previously shut-down Soviet-era military posts, including 13 air bases, 10 radar stations that allow Moscow to track vessels and aircraft in the region, 20 border outposts, and 10 integrated emergency rescue stations.[13]

Militarization and development in the Arctic is a defining characteristic of Russian domestic and foreign politics, while the Arctic has not, in recent history, drawn as much attention from policymakers in Washington. The United States' indifference to the Arctic on a national policy level is largely due to a domestic political debate between Republicans and Democrats on the existence of climate change and U.S. defense priorities focusing on combating global extremism in the Middle East and Africa for the last two decades. However, the United States declared its preeminent defense and national security concern to be a return to great power competition in the 2018 National Defense Strategy. The renewed interest in great power competition, also referred to as "strategic competition" by the Biden administration, has fueled an interest in competing with Russia in the Arctic.

To effectively compete with Russia in the Arctic, the United States has begun its own remilitarization of the region. In recent years, the United States has explored the construction of a strategic port in the Arctic, invested in ice breakers needed to navigate Arctic waters, and reactivated the U.S. 2nd Fleet, responsible for U.S. Navy activity on the East Coast and North Atlantic, including the Arctic. In

---

12 Ernie Regehr, "Military Infrastructure and Strategic Capabilities: Russia's Artic Defense Posture," in *The Arctic And World Order*, 192–93.

13 Melino and Conley, "The Ice Curtain."

May 2019, then U.S. secretary of state Mike Pompeo asserted that the United States was engaged in strategic military competition with Russia in the Arctic when he stated, "On the security side, partly in response to Russia's destabilizing activities, we are hosting military exercises, strengthening our force presence, rebuilding our icebreaker fleet, expanding Coast Guard funding, and creating a new senior military post for Arctic Affairs."[14]

ARCTIC FLASHPOINTS: A CRISIS IN WAITING?

With Washington and Moscow both beefing up their military presence in the Arctic, the chances of a crisis developing between the two rivals may increase as the sea ice diminishes. When it comes to security challenges, climate change is often described as a "threat multiplier" because it is highly unlikely climate change-related effects will ever directly lead to conflict (i.e., the United States and Russia will not clash over the lack of sea ice in the Arctic), but it makes pre-existing threats or tensions far more dangerous. In an increasingly warmer Arctic, there are several possible flashpoints that could lead to a crisis, including a miscalculation over a U.S. freedom of navigation operation (FONOP) through the region and an accidental military run in between uniformed forces from the United States or NATO and Russia.

The first Arctic flashpoint that could lead to a crisis is Washington and Moscow's differing views over what constitutes a state's territorial waters. Such ambiguity could lead to a miscalculation over a FONOP. In January 1985, the Soviet Union established a system of strait baselines that enclosed many of the bays, estuaries, and navigation straits along the Soviet Arctic coast. The system deemed the bays, estuaries, and navigation straits internal waters of the Soviet Union where Moscow retained complete sovereign control.[15] In spite of the Soviet Union's collapse more than three decades ago, the same navigation routes through the NSR remain closed today by the strait baselines created in 1985. But the legal status of the navigation straits remains hotly contested, particularly over the role that the right of innocent passage, codified in the UN Convention for the Law of the Sea, plays in navigating the NSR, among other disputes. During the last quarter century, Russia has also issued numerous domestic laws, including the 1998 Federal Act on Internal Maritime Waters, Territorial Sea, and Contiguous Zone of the Russian Federation, which states that the 12-nautical mile territorial sea limit also applies to islands of the Russian Federation, and the 2012 Federal Law, which established the guidelines for foreign-flagged vessels entering the NSR.[16] Most recently, the Russian Federation passed the 2013 Rules of Navigation of the Water Area of the Northern Sea Route and updated its extensive regulatory system for navigating the NSR in 2017, which shape how Russia views the international status, or lack thereof, of the NSR. The U.S. government's view of the NSR differs greatly from Russia's. The divergent views are best demonstrated by the May 2015 diplomatic note the United States sent to Russia regarding its regulatory scheme for the Arctic waterway. The note laid out issues Washington has with Russia's governance of the waterway, including pointing out inconsistencies with international law such as requiring Russia's consent to enter the country's exclusive economic zone and territorial sea and Russia's characterization of international straits that form part of the NSR as internal waters.[17]

14   Blake Hounshell, "Pompeo aims to counter China's ambitions in the Arctic," *Politico*, May 6, 2019, https://www.politico.com/story/2019/05/06/pompeo-arctic-china-russia-1302649.
15   Lawson W.Q. Brigham, "Governance and Economic Challenges for the Global Shipping Enterprise in a Seasonally Ice-Covered Arctic Ocean," in *The Arctic And World Order*, 151.
16   "Maritime Claims Reference Manual: Russian Federation 2021," U.S. Navy Judge Advocate General's Corps, March 2021, https://www.jag.navy.mil/organization/code_10_mcrm.htm.
17   J. Ashley Roach, "Freedom of the Seas in the Arctic Region," in *The Arctic And World Order*, 233.

In response to the dispute, the United States could handle the situation similarly to how it handles maritime territorial disputes with China in the South China Sea, including conducting a FONOP. Conducting a FONOP in the Arctic, through the NSR, would reinforce the U.S. position that the NSR, similar to the South China Sea, is international waters. According to the U.S. Department of Defense, a FONOP "includes more than the mere freedom of commercial vessels to transit through international waterways.... The Department uses 'freedom of the seas' to mean all of the rights, freedoms, and lawful uses of the sea and airspace, including for military ships and aircraft."[18] In 2019, U.S. secretary of the Navy Richard V. Spencer spoke about having the Navy send vessels to transit the Arctic in a FONOP, and the sentiment that a FONOP in the Arctic would be directed toward Russia was reinforced when General Curtis Scaparrotti, former supreme allied commander in Europe, testified to the idea in front of the Senate Armed Services Committee.[19] While the United States has not followed through on its statements regarding conducting a FONOP in the Arctic, doing so would be provocative and highly risky because it is uncertain how Russia may choose to respond. If Russia did choose to respond aggressively, it would likely lead to Russia directly confronting Navy ships, which could quickly spiral into a full-blown conflict, especially if U.S.-Russian relations are already tense.

Figure 3: NATO Ships Participating in Cold Response 2022

Source: "NATO Ships Complete Exercise Cold Response," U.S. Navy, April 6, 2022, https://www.navy.mil/Press-Office/News-Stories/Article/2990779/nato-ships-complete-exercise-cold-response/.

The second Arctic flashpoint that could lead to a U.S.-Russian crisis is a direct military encounter between uniformed members of the U.S./NATO and Russian militaries stationed in the Arctic.

---

18   U.S. Department of Defense, *Asia-Pacific Maritime Security Strategy* (Washington, DC: 2015), 2, https://dod.defense.gov/Portals/1/Documents/pubs/NDAA%20A-P_Maritime_SecuritY_Strategy-08142015-1300-FINALFORMAT.PDF.

19   Malte Humpert, "U.S. Navy Plans to Send Surface Vessels Through Arctic," High North News, March 11, 2019, https://www.highnorthnews.com/en/us-navy-plans-send-surface-vessels-through-arctic; and David Auerswald, "Now is not the time for a FONOP in the Arctic," War on the Rocks, October 11, 2019, https://warontherocks.com/2019/10/now-is-not-the-time-for-a-fonop-in-the-arctic/.

Should a confrontation occur, one of the most likely sources would be the Navy's 2nd Fleet or a NATO exercise. The reactivation of the 2nd Fleet means there is a uniformed U.S. military presence responsible for overseeing the Arctic. In addition, the U.S. Coast Guard also has an Arctic presence as the U.S. entity responsible for operating ice breakers, which are critical for many maritime expeditions in the icy Arctic waters. Similarly, 5 of the 30 NATO member states are Arctic states and have strong interests in the region.[20] At the time of this writing, the United States, Canada, Denmark, Iceland, and Norway are members of the military alliance, while two additional Arctic states, Sweden and Finland, have been granted entry into NATO (pending domestic legislature approval in both countries) in response to the Russian invasion of Ukraine in February 2022. NATO, and its soon to be expanded roster of Arctic states, is demonstrating its commitment to the Arctic through the running of military exercises in the region, most recently conducting Cold Response 2022, an exercise in March 2022 designed to bring together thousands of NATO troops and test the ability of the alliance to operate in cold weather conditions across Norway. Similarly, in October 2018, NATO conducted the military exercise Trident Juncture off the coast of Norway. The exercise is especially noteworthy as it featured the U.S. aircraft carrier *Harry S. Truman* and its strike group. The inclusion of the *Truman* and its strike group made them the first U.S. surface vessels to travel into the Arctic in three decades. With such a heavy NATO presence in the Arctic and tensions high in Europe as a result of the 2022 Russian invasion of Ukraine, an accidental confrontation between NATO members and Russia in the Arctic could trigger Article V of the North Atlantic Treaty, the mutual defense article, inviting a full-blown NATO-Russia war.

CRISES, ESCALATION MANAGEMENT, AND RUSSIAN NUCLEAR WEAPONS POLICY

Should a crisis break out between the United States and Russia, both sides should endeavor to refrain from escalating the crisis to avoid further conflict, most critically by refraining from using nuclear weapons. Successful crisis management, especially of a crisis that involves two nuclear powers and the chance of nuclear escalation, involves successfully contending with additional stresses on a country's government. Andrei Kokoshin, former secretary of the Russian Security Council, suggests that managing a crisis "requires special concentration and will from the leaders; special mechanisms for the collection, processing and reporting of information to the leadership; special control of the activities of the country's forces; special instructions for ambassadors and military attachés; a special regime of communication with the national and the foreign mass media, and many other things."[21]

One complication in modern crisis management is the evolution of information sharing. During the Cold War, the speed at which information proliferated during a crisis was far slower than in 2022, where social media applications such as Twitter and Facebook allow for near instantaneous information sharing, whether true or not. Following the Russian invasion of Ukraine, the prospects for either the U.S. or Russian governments successfully managing an additional crisis is suspect at best. In the United States, political polarization is arguably at an all-time high, which often leads to gridlock, finger pointing, and inaction in Washington. The political environment in the United States is further complicated by the information environment, including the presence of "fake news," most notably highlighted by the prominence of conspiracy theories such as QAnon. In Russia, the invasion

---

20   At the time of this writing, NATO's official membership does not include Sweden and Finland.

21   Anya Loukianova Fink, "Crisis Stability in the Twenty-First Century: Discussion Paper for a Track II Dialogue on U.S.-Russian Crisis Stability," CSIS, 2018, https://csis-website-prod.s3.amazonaws.com/s3fs-public/180508_Fink_CrisisStability.pdf?vLZ3_YNwaUGjWNGxgoBUb7D6mBKHplp2.

of Ukraine has shown how information breakdowns within the highest levels of Russia's military leadership can happen in a high-stress situation; U.S. intelligence reported that there has been a "clear breakdown in the flow of accurate information" between Putin and his advisers and that Putin has been misinformed on his military's performance in Ukraine because advisers are "scared to tell him the truth."[22]

Should the United States and Russia face a crisis over a FONOP or a military confrontation in the Arctic, the Kremlin will turn to its escalation management strategy, strategic deterrence. The Russian strategy of strategic deterrence, defined by their 2015 National Security Strategy, is "a series of interrelated political, military, military-technical, diplomatic, economic, and informational measures to prevent the use of force against Russia, defend sovereignty, and preserve territorial integrity."[23] Russian strategic deterrence is continuously in use. During peacetime, strategic deterrence is used to contain and deter enemies from using force against Russia, but in wartime, or during a crisis, it is used to manage escalation. While Russian strategic deterrence includes non-military measures, such as of a political, diplomatic, legal, economic, ideological, and technical-scientific nature, Russian views on deterrence foremost revolve around the use of military power as a tool of coercion. For example, Russian military measures could include demonstrating the power of the military, raising readiness, deploying forces, and conducting or threatening to conduct military strikes. In total, Russia's strategic deterrence strategy seeks to ultimately raise the adversary's expected costs above the desired benefits. One-way strategic deterrence can be achieved is through threatening to use, or *actually* using, nuclear weapons as a means to raise the adversary's costs above the desired benefits.

Russia is estimated to possess a largely modernized nuclear weapons arsenal—with 4,447 deployed or reserve warheads—making it the largest in the world, with ~600 more warheads than the United States.[24] Russian nuclear force posture includes about 1,588 strategic warheads that would be delivered via ballistic missile or strategic bomber, while ~977 strategic warheads and ~1,912 non-strategic warheads are held in reserve.."[25]

With the largest nuclear arsenal in the world, Russia also has a stated plan and criteria that could trigger the use of the weapons. In June 2020, the *Basic Principles of the Russian Federation's State Policy in the Domain of Nuclear Deterrence* was resealed, which is the document that lays out the two conditions in which Russia would use nuclear weapons. The document states, "The Russian Federation retains the right to use nuclear weapons in response to the use of nuclear weapons and other types of weapons of mass destruction against it and/or its allies and also in the case of aggression against the Russian Federation with the use of conventional weapons, when the very existence of the state is put under threat."[26] While the first half of the policy declaration is not

22   Nicholas Reimann, "Putin's Advisors Misleading Him About Russian Performance In Ukraine, U.S. Official Says," *Forbes*, March 30, 2022, https://www.forbes.com/sites/nicholasreimann/2022/03/30/putins-advisors-misleading-him-about-russian-performance-in-ukraine-us-official-says/?sh=5d3d12e7394a; and Aamer Madhani and Nomaan Merchant, "White House: Intel shows Putin misled by advisers on Ukraine," AP News, March 30, 2022, https://apnews.com/article/russia-ukraine-putin-europe-00716c99579afeff701af31b32ef7c8c.

23   Michael Kofmna and Anya Loukianova Fink, "Escalation Management and Nuclear Employment in Russian Military Strategy," War on the Rocks, June 23, 2020, https://warontherocks.com/2020/06/escalation-management-and-nuclear-employment-in-russian-military-strategy/.

24   Hans M. Kristensen and Matt Korda, "Nuclear Notebook: How many nuclear weapons does Russia have in 2022?," Bulletin of the Atomic Scientists, February 23, 2022, https://thebulletin.org/premium/2022-02/nuclear-notebook-how-many-nuclear-weapons-does-russia-have-in-2022/.

25   Ibid.

26   David Halloway, "Read the fine print: Russia's nuclear weapon use policy," Bulletin of the Atomic Scientists, March 10, 2022, https://thebulletin.org/2022/03/read-the-fine-print-russias-nuclear-weapon-use-policy/.

surprising, the second half—"Also in the case of aggression against the Russian Federation with the use of conventional weapons, when the very existence of the state is put under threat"—is arguably the foundation for a limited nuclear weapons use policy that could lead to further use of nuclear weapons. While a U.S.-Russian crisis in the Arctic would not likely meet the threshold for use of nuclear weapons by either definition put forth, one factor that could complicate the matter is the status of Russia and its perceived vulnerability when the 2022 Russian-Ukrainian war eventually reaches its end point. Most troubling would be if the war results in Russia failing to achieve its goals while taking heavy military and economic losses, leaving it internationally isolated and backed into a corner. Should that be the case, Russia may revert back to its 1990s deterrence posture of relying solely on nuclear weapons to deter conflict, which would increase the likelihood of a nuclear incident or, worse, leave the country feeling the need to lash out and escalate a crisis, potentially through using limited nuclear strike in the Arctic to deter U.S. actions and preserve some level of its national pride and status.

THE RETURN OF DUCK AND COVER?

While climate change's effects in the Arctic are quite visible and create the opportunity for potential flashpoints between the United States or NATO and Russia, the likelihood that a crisis would break out over a U.S. FONOP through the Arctic or a military confrontation remains quite low. Recent history shows that while FONOPs can be provocative and risky, conducting such operations has not led to an escalation of tensions between two competitive states. For example, in 2018 and 2019, during a time at which U.S.-China relations were deteriorating, the U.S. Navy sailed vessels more than a dozen times within 12 nautical miles of territory China claims in the South China Sea, none of which led to any further escalation other than strongly worded statements from political officials.[27] While FONOPs remain a powerful method of signaling Washington's commitment to freedom of the seas worldwide and play a role in achieving larger foreign policy goals, the exercises themselves have not sparked any crises or conflicts to date.

Similarly, there is no need to dust off the duck and cover drill instructions when considering if a confrontation between U.S./NATO and Russian militaries would lead to a nuclear conflict. U.S. and NATO military exercises in the Arctic have occurred for more than a decade, including two prominent exercise series, Trident Juncture and Cold Response, that have been taking place at regular intervals since 2015 and 2006, respectively. Additionally, the United States and Canada have previously conducted Arctic exercises, such as Operation Noble Defender, a North American Aerospace Defense Command operation that seeks to "demonstrate the ability to integrate with other defense and security partners for a holistic defense of North America," including in the Arctic.[28] Arctic exercises and military drills are not just a uniquely U.S. and NATO staple, as Russia has also included Arctic forces in its series of regularly scheduled military exercises. Zapad, an exercise series designed for the Western Military District of Russia, includes the Northern Fleet tasked with operating in the Arctic and runs every four years, the most recent iteration occurring in 2021. In addition to major regional military exercises, Russian miliary training also features smaller, more focused exercises. For example, an exercise in the Barents Sea in January 2022 included 30 warships, 20 aircraft, and 1,200

27   Oriana Skylar Mastro, "Military Confrontation in the South China Sea," Council on Foreign Relations, *Contingency Planning Memorandum*, no. 36, May 21, 2020, https://www.cfr.org/report/military-confrontation-south-china-sea.

28   "Operation Noble Defender," North American Aerospace Defense Command, February 8, 2022, https://www.norad.mil/Newsroom/Fact-Sheets/Article-View/Article/2928028/operation-noble-defender/.

personnel and was designed to show Arctic readiness and Moscow's ability to protect the NSR.[29] With an increased military presence in the region from both the United States (and NATO) and Russia, the worry that the two militaries could confront each other or accidentally run into each other, thus sparking a crisis, increases. However, U.S. and Russian military collisions—including instances where vessels from the two states' navies literally collide with each other—are also unlikely to spiral into a crisis. Despite the strong military presence and drills in the Arctic, exercises have come and gone for years without leading to any major incidents, and the United States and Russia have a history of keeping tensions low when their navies literally collide.

In the rare case of an actual collision, there are also multiple examples of U.S.-Soviet run-ins during the Cold War that were contained and, in one case, not even made public until more than 40 years later. As part of the ongoing superpower rivalry, the U.S. and Soviet navies often operated in close proximity to one another and routinely played cat-and-mouse games with each other, which led to incidental collisions. In 1974, a Soviet attack submarine struck the *James Madison*, a U.S. Navy ballistic missile submarine, while both of the nuclear-powered vessels were operating off the coast of Scotland.[30] Similarly, 10 years later, a Soviet attack submarine struck the U.S. Navy aircraft carrier USS *Kitty Hawk* in the Sea of Japan.[31] The *Kitty Hawk* sustained no damage itself, but the Soviet submarine was dead in the water and required assistance from a Soviet cruiser. In the immediate aftermath of the collision, both sides acted in accordance with the 1972 Incidents-At-Sea Treaty, a law that stipulates neither U.S. nor Soviet navies will interfere with the operations of the other, nor will they operate in ways that would be viewed as threatening. In both circumstances, despite literal collisions between adversarial navy vessels, the incidents did not lead to escalation.

CONCLUSION

The effects of climate change in the Arctic, seen largely through melting sea ice that allows Arctic waterways to be navigable for longer stretches of the year, is playing a major role in the ongoing strategic competition between the United States and Russia. While the diminishing levels of sea ice and strategic competition can increase the likelihood of a crisis breaking out that could spiral into conflict, and even nuclear war, the likelihood of that happening is very low. If a crisis should breakout, there are two likely catalysts: (1) a U.S. FONOP through the Artic designed to reinforce its belief that Arctic waterways, especially the NSR, are international waterways, or (2) an accidental run-in between the U.S./NATO and Russian militaries operating in the Arctic due to the high levels of military activity in the region. While both flashpoints create an opportunity for a crisis to break out, recent history shows that U.S. FONOPs directed against adversarial states and military exercises, and even the literal colliding of adversarial ships, are unlikely to ignite a crisis.

29   Gabrielle Tétrault-Farber, "Russia starts navy drills to rehearse protecting Arctic shipping lane," Reuters, January 26, 2022, https://www.reuters.com/world/europe/russia-starts-navy-drills-rehearse-protecting-arctic-shipping-lane-2022-01-26/.

30   David Axe, "When a Soviet Submarine Collided with this U.S. Navy Aircraft Carrier," *The National Interest*, December 17, 2021, https://nationalinterest.org/blog/reboot/when-soviet-submarine-collided-us-navy-aircraft-carrier-198081.

31   Richard Holloran, "Soviet Sub and U.S. Ship Collide," *New York Times*, March 22, 1984, https://www.nytimes.com/1984/03/22/world/soviet-sub-and-us-ship-collide.html.

Small Modular Reactors (SMRs) and Their Implications for International Nuclear Supply Agreements

By Aubrey Means[1]

INTRODUCTION

The peaceful use of nuclear power has the potential to significantly contribute to combating climate change as a reliable source of electricity with zero carbon emissions. However, legitimate concerns over the safety, security, and proliferation of nuclear material and technology must be addressed as the world's economies increasingly rely on nuclear energy. In short, how can the world harness nuclear energy while mitigating the technology's attendant risks?

Addressing this question raises two main challenges. First, how can the world reduce the competitive cost of nuclear energy production and make it more accessible to countries around the world? The enormous financial and infrastructural investments required to build a traditional nuclear power plant are prohibitively expensive for many would-be nuclear newcomers. Emerging economies today bear the additional burden of having to achieve economic development without relying on the heavy carbon-emitting technologies used by the economic giants of the twentieth century. Meanwhile, in order to reduce carbon emissions to the threshold recommended by the Intergovernmental Panel on

---

1   Aubrey Means is a national security specialist at Pacific Northwest National Laboratory. The views expressed in this paper are her own and do not necessarily reflect the views of her employer.

Climate Change (IPCC), nuclear power must make up a greater share of global energy sources than it currently does. The spread of nuclear power generation to more developing countries would address their domestic energy production needs as well as the global need to reduce carbon emissions. This necessitates a more affordable, scalable means of entry into the nuclear power "club."

The second challenge has to do with nuclear safety, security, and nonproliferation. In order to encourage investment in nuclear technology, policymakers and legislators must be convinced that its spread will neither result in massive radiation disasters, nor be co-opted by political motivations and turned into a covert nuclear weapons program. Therefore, both existing and cutting-edge nuclear facilities must implement technology and practices which are safe (i.e., preventing radiation accidents) and secure (i.e., preventing malign actors from weaponizing material).

The challenge of making nuclear power more accessible to newcomer countries without creating additional environmental or security problems must be addressed—and soon, according to climate science. This article will examine the role small modular reactors (SMRs) can play as part of a solution and the ways in which nuclear supply states can bolster their supply agreements to enhance the security of the nuclear industry.

NUCLEAR POWER AND CLIMATE CHANGE

It is first important to recognize why nuclear power must play an integral role in combating climate change. The environmental effects of climate change are not only a concern for populations around the world and their relationship with natural resources, but they also impact national security. According to a 2019 congressional study, 23 of the 79 U.S. military bases threatened by extreme climate effects are related to the U.S. nuclear mission, and seven of those sites house a total of nearly 6,000 nuclear warheads.[2] Furthermore, increased national unrest in those countries most vulnerable to the immediate effects of climate change, such as Syria, poses a security threat to the United States and other Western powers, expressed through issues such as terrorist activity and mass migrations. The United States, as one of the world's largest carbon emitters, has a vested interest in preventing climate catastrophe.

In 2018, on the basis of current levels of carbon emissions, the IPCC predicted that the earth is due to surpass a critical climate threshold by 2040, at which point the effects of climate change would become irreversible.[3] More recently, in February 2022, the IPCC produced a new assessment which argued that current efforts at curbing carbon emissions are not sufficient to avoid that threshold.[4] Nuclear power, which produces zero-carbon emissions and generates a lower cost per unit of electricity than wind or solar power, must therefore be an integral part of any climate strategy. Renewables will play a role as well, but at present they are not capable of replacing fossil fuels at the scale and rate necessary to reverse global warming. Technological breakthroughs that could shift this dynamic remain elusive—despite billions of dollars invested.[5] Energy technology has a

---

2   Matt Korda, "The US Nuclear Deterrent Is Not Prepared for Climate Crisis," *Forbes*, March 16, 2020, https://www.forbes.com/sites/matthewkorda/2020/03/16/the-us-nuclear-deterrent-is-not-prepared-for-climate-catastrophe/#6d83042d3a1e.

3   Intergovernmental Panel on Climate Change, *Special Report: Global Warming of 1.5°C* (Geneva: October 2018), https://www.ipcc.ch/sr15/.

4   Intergovernmental Panel on Climate Change, *Climate Change 2022: Impacts, Adaptation and Vulnerability* (Geneva: February 2022), https://www.ipcc.ch/report/ar6/wg2/.

5   Seth Kirshenberg, Richard Butterworth, Hillary Jackler, and Brian Oakley, *Examination of Federal Financial Assistance in the Renewable Energy Market* (Washington, DC: U.S. Department of Energy, October 2018), https://www.energy.gov/sites/default/files/2018/11/f57/Examination%20of%20Federal%20Financial%20Assistance%20in%20the%20Renewable%20Energy%20Mark..._1.pdf.

history of incremental improvement; the world cannot rely on technological miracles to meet the IPCC's approaching deadline when it has proven and reliable—as well as new—nuclear technology already at its disposal. This article will not address the parallel contributions of renewable energy technology, therefore, and will assume that nuclear technology will become more widespread globally as an indispensable component of any climate strategy.

Unlike solar and wind power, which depend on weather patterns to produce energy and therefore only run for an average of 24.5 percent and 34.8 percent of their respective total installed capacities, nuclear plants provide a nearly constant baseload of power, running at an average of 93.5 percent of their total capacity in the United States.[6] This means that a large city such as New York or Tokyo can support industries and resident populations even when the sun is not shining or the wind is not blowing. In fact, renewable energy sources are most efficient when they are used in combination with a more constant source of energy, such as nuclear, so that the supply of energy always matches demand. Nuclear power also has a much smaller impact on land use: the Department of Energy calculates that roughly 431 wind turbines or more than 3 million solar panels would be required to produce the same amount of electricity as a single nuclear power plant.[7] All of those solar panels will also eventually generate millions of metric tons of waste when they wear out, given that the industry standard lifespan of a solar panel is about 25 years.[8] Compare that to nuclear power plants; those first constructed in the 1960s initially had a lifespan between 40 and 60 years, and many have been extended to almost 80 years.

Of course, nuclear power generates its own waste, which is often the focus of anti-nuclear movements in the United States and around the world. However, it is important to consider nuclear waste in terms of its scale: all of the spent nuclear fuel produced in the United States since the beginning of the nuclear industry amounts to about 60 dry storage casks, enough to fit on a single football field within the 10-yard line.[9] This spent fuel is currently stored across 34 states. Long-term storage solutions for nuclear waste are in development around the world, some of the most promising being the deep geological repositories proposed in Sweden and Finland, which are currently undergoing licensure.[10] Such solutions would enable countries around the world to produce nuclear energy long into the future without having to worry about what to do with the material once it has been used.

Climate change has driven a resurgent interest in nuclear energy around the world. Some of the main takeaways from the UN Climate Change Conference in Glasgow (COP26), the iteration of the UN Climate Change Conference hosted by Scotland in 2021, included new nuclear initiatives announced by China and several European states to supplement and support their renewable energy strategies.[11] The United Kingdom released a new national energy strategy in April 2022 which details investments of over £2 billion ($2.4 billion) in new nuclear technologies,

---

6   "What is Generation Capacity?," U.S. Department of Energy, Office of Nuclear Energy, May 1, 2020, https://www.energy.gov/ne/articles/what-generation-capacity.

7   "The Ultimate Fast Fact Guide to Nuclear Energy," U.S. Department of Energy, Office of Nuclear Energy, January 2019, https://www.energy.gov/sites/prod/files/2019/01/f58/Ultimate%20Fast%20Facts%20Guide-PRINT.pdf.

8   Nate Berg, "What Will Happen to Solar Panels After Their Useful Lives Are Over?," GreenBiz, May 11, 2018, https://www.greenbiz.com/article/what-will-happen-solar-panels-after-their-useful-lives-are-over.

9   Luisa Kenausis, "Nuclear Waste Issues in the United States," Arms Control Center, August 22, 2018, https://armscontrolcenter.org/nuclear-waste-issues-in-the-united-states/.

10   Cindy Vestergaard, Rowen Price, and Trinh Le, "Geological Disposal of Spent Nuclear Fuel," Stimson Center, September 2021, https://www.stimson.org/2021/geological-disposal-and-spent-nuclear-fuel/.

11   Emma Derr, "COP26: Nuclear Plays a Central Role," Nuclear Energy Institute, November 16, 2021, https://www.nei.org/news/2021/cop26-nuclear-plays-a-central-role.

including £210 million ($252 million) specifically dedicated to SMRs.[12] In November 2021, the Biden administration signed into law the Infrastructure Investment and Jobs Act, which provides nearly $8 billion to support the existing fleet of U.S. nuclear plants as well as to further develop advanced reactors, including SMRs.[13] Clearly, the world is reconsidering nuclear power's potential to contribute to a climate change solution.

In addition to the climate context, however, Russia's February 2022 invasion of Ukraine has provided yet another motivation to invest in nuclear power: energy independence. Russia's actions, horrifying on both a political and humanitarian level, also have implications for countries within Europe and beyond who rely on Russian hydrocarbon exports to supply their national energy grids. No state wants to be dependent on an aggressive superpower without respect for national sovereignty; the threat of cutting off its target's energy supply is an extremely powerful tool in Russia's tool kit. Developing the infrastructure and technology to produce a reliable source of energy within a country's own borders, therefore, is another newly compelling reason to invest in nuclear power. Additionally, Russia's military usurpation and aggression surrounding the Zaporizhzhia nuclear power plant in eastern Ukraine illustrate some of the security concerns which must be addressed at any nuclear facility, which will be explored later in this article.

SMALL MODULAR REACTORS

Dozens of new start-up companies aim to make investing in nuclear power simpler and safer through advanced nuclear technology. One of the most intriguing applications of this new technology comes in the form of SMRs, which are designed to be scalable versions of more traditional power plants for a fraction of the financial, material, or infrastructural investment. Whereas a traditional pressurized water reactor (PWR) may produce about one gigawatt (GW) of electricity, SMRs—which can include miniaturized PWR designs as well as alternative advanced reactor designs—typically produce less than 300 megawatts (MW) of electricity. Particularly appropriate for economies in the developing world, the modularity of SMRs would theoretically allow for a country to begin with one or two modules and then scale up to greater power generation as industry expands.

SMRs are also cheaper than building traditional nuclear power plants. Most of the competitive cost associated with producing nuclear energy is tied up in the siting and construction of the plant itself, which must be customized each time and involves significant indirect costs of labor, office engineering, and quality control. A 2015 study of the costs of electricity generation found that a 10 percent increase in construction time of a traditional nuclear power plant led to cost increases of between 18 and 22 percent.[14] In contrast, the relatively small size and standardized design of SMRs means that they can be mass produced in a factory and shipped to their ultimate destination. In an analysis of successful nuclear energy deployments around the world between 1966 and 2002, the regions where standardization was applied showed a clear reduction in construction time and overhead costs.[15] SMRs open new locations to reactor siting and dramatically lower costs.

12   "British Energy Security Strategy," Government of the United Kingdom, April 7, 2022, https://www.gov.uk/government/publications/british-energy-security-strategy/british-energy-security-strategy.

13   U.S. Congress, House, "Infrastructure Investment and Jobs Act," H.R. 3684, 117th Cong., 1st sess., Introduced June 6, 2021, https://www.congress.gov/bill/117th-congress/house-bill/3684/text.

14   OECD Nuclear Energy Agency and International Energy Agency, *Projected Costs of Generating Electricity: 2015 Edition* (Paris: OECD Nuclear Energy Agency and International Energy Agency, October 2015), https://www.iea.org/reports/projected-costs-of-generating-electricity-2015.

15   M. Berthelémy and L. Rangel, "Nuclear Reactors' Construction Costs: The Role of Lead-Time, Standardization, and Technical Progress," *Energy Policy* 82 (March 2015): 118–130, doi:10.1016/j.enpol.2015.03.015.

The size and flexibility of SMRs also mean that they can be used for nontraditional power applications. Rosatom, Russia's state-owned nuclear power company, has recently signed a deal with a Russian mineral extraction corporation, Seligdar, to further develop gold deposits in the Yakutia region using an RITM-200N SMR to provide sufficient power for mining operations in such a remote area.[16] Additionally, the International Atomic Energy Agency (IAEA) recently celebrated a major victory over TR4, a persistent pathogen that destroys crops, using nuclear technology which could easily be driven by SMRs. In 2021, the disease affected crops across Peru and Colombia, threatening about 47 percent of the world's banana production, with major impacts on South American economies. In partnership with the UN Food and Agricultural Organization, the IAEA used irradiation to develop disease-resistant varieties of plants and stopped the spread of TR4.[17] SMRs could make these and other beneficial applications such as hydrogen production or seawater desalination, in addition to traditional electricity production, more accessible for countries around the world.

Indeed, the private sector has already started targeting real-world applications for SMRs. NuScale Power, an Oregon-based start-up which was the first to have its SMR design approved by the Nuclear Regulatory Commission (NRC), has signed multiple memorandums of understanding (MOUs) to explore the deployment of its SMR plants—nicknamed VOYGR—around the world. Most recently, an MOU signed with Kazakhstan Nuclear Power Plants LLP (KNPP) in December 2021 will enable NuScale to assess plant engineering, construction, commissioning, operation and maintenance, and project-specific studies. Though Kazakhstan is the world's largest producer of uranium, it currently has no operational nuclear facilities. In order to reach their declared goal of attaining carbon neutrality by 2060, KNPP has determined that "small modular reactors are the most promising for Kazakhstan … for the country's transition to a green economy."[18]

NuScale is also invested in SMR energy production closer to home through its Carbon Free Power Project (CFPP), an initiative "to provide safe, reliable, and cost-competitive clean energy to communities across the mountain West," beginning in Idaho Falls, Idaho. The project is in partnership with the Utah Associated Municipal Power System (UAMPS), Idaho National Lab (INL), and Energy Northwest and aims to begin generating power in 2029 or 2030.[19] In addition to providing clean energy, the project is expected to benefit local and state economies by creating some 1,600 jobs over the course of plant construction and an additional 600 to 700 jobs every year that the plant is in operation.[20] If successful, the CFPP and its unique ownership model would set a precedent for the use of SMRs in partnership with established utility companies to reinvent a community or region's power supply on a carbon-free basis, which may be likewise implemented in other states and around the world.

In another instance of near-term implementation of SMR technology, GE Hitachi Nuclear Energy, an alliance between General Electric in the United States and Hitachi in Japan, announced in December 2021 their intention to cooperate with the Polish firm Synthos Green Energy to deploy at least

---

16   "Deal Signed for Nuclear to Power Russian Gold Mine," World Nuclear News, January 21, 2022, https://www.world-nucle-ar-news.org/Articles/Deal-signed-for-nuclear-to-power-Russian-gold-mine.

17   "IAEA Combats Crop-Threatening Banana Wilt with Nuclear Technology," Nuclear Newswire, American Nuclear Society, January 6, 2022, https://www.ans.org/news/article-3555/iaea-combats-cropthreatening-banana-wilt-with-nuclear-tech-nology/.

18   "NuScale Pondering SMRs for Kazakhstan," American Nuclear Society, Nuclear News, December 20, 2021, https://www.ans.org/news/article-3525/nuscale-pondering-smrs-for-kazakhstan/.

19   "Carbon Free Power Project," NuScale Power, https://www.nuscalepower.com/projects/carbon-free-power-project.

20   Geoffrey Black and Steven Petereson, "Economic Impact Report: Construction and Operation of a Small Modular Reactor Electric Power Generation Facility at the Idaho National Laboratory Site, Butte County, Idaho," Idaho Policy Institute, Boise State University, January 29, 2019, https://www.nuscalepower.com/projects/carbon-free-power-project.

10 BWRX-300 SMRs in Poland by 2030.[21] The agreement also involves supply chain support from Ontario-based BWXT Canada, highlighting just how the field is growing globally. Similar projects utilizing SMR designs are emerging in South Korea, Russia, the United Kingdom, and in several other European countries.

To add momentum to the growing interest in SMR technology, the Biden administration recently introduced a new initiative: the Foundational Infrastructure to Support Small Modular Reactor Technology (FIRST) program. Specifically, the FIRST program seeks to "provide capacity-building support for partner countries as they develop their nuclear energy programs [for] clean energy goals under the highest international standards for nuclear safety, security, and nonproliferation."[22] In February 2022, the United States, Japan, and Ghana launched a new initiative under the FIRST program to support stakeholder engagement, technical collaboration, and logistical planning for Ghana's adoption of SMR technology.[23] In January, Estonia also agreed to cooperate with the United States by utilizing FIRST's resources in government, academic, and technical expertise to explore the feasibility of transitioning from carbon-intensive power sources to SMR technology.[24] Furthermore, the United States signed a brand-new nuclear cooperation memorandum of understanding (NCMOU) with Armenia in May for similar pursuits.[25]

Government initiatives to encourage nuclear development are increasing. In addition to the FIRST program under the Department of State, the Department of Energy's Advanced Reactor Demonstration Program aims to promote the deployment of reactors with various advanced technology designs. These range from remote or mobile applications of a few million watts of electricity (Mwe) to modular designs intended to be scaled up to a few hundred MWe. Congressional support also appears to be growing stronger. In December 2021, Representatives Anthony Gonzalez (R-OH) and Elaine Luria (D-NY) introduced the bipartisan Accelerating Nuclear Innovation Through Fee Reform Act (H.R. 6154), which would encourage private sector investment in advanced nuclear technologies by reducing fees associated with NRC design approval.[26] Additionally, from a nuclear supply perspective, the bipartisan International Nuclear Energy Act (S. 4064) was introduced to the Senate Committee on Foreign Relations in April 2022 with an intent to promote further engagement with partners, such as Ghana, Estonia, and Armenia, and to develop a civil nuclear export strategy for the United States which can offset the influence of Russia and China in this sphere.[27] Such legislation will eventually enable more projects such as NuScale's CFPP to benefit regional economies and energy grids across the United States and around the world.

---

21 "Firms Partner to Support BWRX-300 Development in Poland," American Nuclear Society, Nuclear News, December 16, 2021, https://www.ans.org/news/article-3519/firms-partner-to-support-bwrx300-deployment-in-poland/.

22 "Program to Create Pathways to Safe and Secure Nuclear Energy Included in Biden-Harris Administration's Bold Plan to Address the Climate Crisis," U.S. Department of State, Office of the Spokesperson, April 27, 2021, https://www.state.gov/program-to-create-pathways-to-safe-and-secure-nuclear-energy-included-in-biden-harris-administrations-bold-plans-to-address-the-climate-crisis/.

23 "U.S. and Ghana Take Next Steps Forward on Clean, Safe Nuclear Energy," U.S. Embassy in Ghana, March 8, 2022, https://gh.usembassy.gov/u-s-and-ghana-take-next-steps-forward-on-clean-safe-nuclear-energy/.

24 Sten Hankewitz, "Estonia Joins a US Nuclear Programme," Estonia World, January 26, 2022, https://estonianworld.com/technology/estonia-joins-a-us-nuclear-programme/.

25 "The United States of America and the Republic of Armenia Sign a Memorandum of Understanding Concerning Civil Nuclear Cooperation," U.S. State Department, May 2, 2022, https://www.state.gov/the-united-states-of-america-and-the-republic-of-armenia-sign-a-memorandum-of-understanding-concerning-strategic-civil-nuclear-cooperation/.

26 U.S. Congress, House, "Accelerating Nuclear Innovation through Fee Reform Act," H.R. 6154, 117th Cong., 1st sess., Introduced December 7, 2021, https://www.congress.gov/bill/117th-congress/house-bill/6154/text.

27 "Risch, Manchin Introduce International Nuclear Energy Act," Senate Committee on Foreign Relations, April 8, 2022, https://www.foreign.senate.gov/press/ranking/release/risch-manchin-introduce-international-nuclear-energy-act.

SMRs offer access to nuclear power and all its applications, from heat production and electricity generation to medical isotope production, while lessening the cost burden of constructing full-scale nuclear power plants. As the world's first nuclear power and one of its largest carbon emitters, the United States should take a leadership role in modeling and supporting nuclear solutions to the climate crisis around the world. Again, this article assumes that the economic and environmental benefits of advanced nuclear technology will see it increasingly adopted around the world; however, strong legal and regulatory frameworks are required to address associated security concerns. The supply of nuclear material, technology, and expertise must be balanced with protections against nuclear accidents or weapons proliferation.

INTERNATIONAL SUPPLY AGREEMENTS

It is no coincidence that the major nuclear superpowers of the Cold War are also the world's leading nuclear supply states. In addition to the United States and Russia, France, the United Kingdom, and China, as well as non-nuclear weapons states such as South Korea, Germany, and Canada, are all influential members of the Nuclear Suppliers Group (NSG) and have conducted various agreements to supply nuclear material or reactor technology to other states. As climate change intensifies the demand for nuclear energy and more countries enter the civil nuclear space, the number of supply agreements between newcomers and these established nuclear energy producers will increase sharply. As a result, it is important that such agreements maintain universally high standards for nonproliferation, security, and safety concerns. This section will primarily examine and compare the 104 nuclear supply agreements that the United States and Russia entered between 1990 and 2020, drawing in part from previous analysis by Adam Stulberg and Jonathan Darsey in an edited manuscript by the James Martin Center for Nonproliferation Studies.[28]

The supply of nuclear technology is pursuant to Article IV of the Nuclear Nonproliferation Treaty (NPT), which asserts every state's right to the use of peaceful nuclear power. Historically, the commercial competition for supply contracts has generated fears of a "race to the bottom" when it comes to security standards. However, both the United States and the Soviet Union recognized the risks of the uncontrolled spread of nuclear technology when India surprised the world with its 1974 detonation of a nuclear weapon. That event inspired the London Club, the predecessor of today's NSG, which gathered nuclear suppliers for the first time and outlined lists of goods and technologies subject to export controls. Such controls were applied universally to maintain nonproliferation and security standards in the face of commercial competition.

Today, nuclear supply agreements come in different forms, but all incorporate some level of these standards. The United States manages its supply of peaceful nuclear technology or equipment through 123 Agreements, named in reference to Section 123 of the U.S. Atomic Energy Act. As of January 2022, the United States had 23 such agreements in force governing peaceful nuclear cooperation with 47 countries and the IAEA.[29] Increasingly, the United States also maintains NCMOUs, which do not permit the export of nuclear material or equipment but do establish broader strategic relationships between the nuclear regulatory and scientific communities in the United States and

28   Adam N. Stulberg and Jonathan Darsey, "Moving Beyond Self-Restraint: Bilateral Commercial Nuclear Supply and US-Russia Tacit Understanding on Nuclear Security and International Safeguards," in *End of an Era: The United States, Russia, and Nuclear Nonproliferation*, edited by Sarah Bidgood and William C. Porter (Monterey, CA: James Martin Center for Nonproliferation Studies, August 2021), 87–128.

29   "123 Agreements for Peaceful Cooperation," National Nuclear Security Administration, January 10, 2022, https://www.energy.gov/nnsa/123-agreements-peaceful-cooperation.

their counterparts elsewhere. There are currently NCMOUs between the United States and Romania (by far the most advanced relationship between a foreign government and a private U.S. nuclear reactor developer), Poland, Slovenia, Bulgaria, Armenia, and Ghana.[30] As initiatives such as the FIRST program continue to grow and more states seek U.S. assistance in reducing carbon emissions, others are likely to be added. This signifies a changing scope from the more formal 123 Agreements, which are predominantly held between the United States and developed economies, many of which already have some degree of nuclear power in their energy mix. In contrast, nuclear supply agreements made by Russia and China historically include more developing countries and newcomers to nuclear power across Africa, Southeast Asia, and Central and South America.

Russian nuclear supply agreements differ slightly from 123 Agreements, but they are similar in the sense that both types of agreement are legally binding and incorporate standards on a few main areas of concern. These areas include requirements to adopt IAEA safeguards; restrictions on direct transfers of certain nuclear material or equipment; limitations on fuel enrichment and reprocessing; and controls over the transfer of nuclear material, technology, or expertise to third parties outside of the agreement. Both the United States and Russia are consistent in their emphasis on the NPT as the cornerstone of the global nonproliferation regime, and many elements of their respective supply agreements are drawn from that treaty.

When it comes to nonproliferation measures, the United States has issued two "gold standard" agreements: one with the United Arab Emirates in 2008 and another with Taiwan in 2013. These are characterized by consistently tight restrictions across all areas of concern listed above and, in particular, prohibit indigenous enrichment and reprocessing of nuclear fuel. Meanwhile, Russia has taken its own approach through unique "take-back" provisions, in which spent fuel produced in a partner country is transported back to Russia for reprocessing and plutonium separation before the waste is returned to the country of origin. Furthermore, in the majority of agreements Russia has signed since 2008, it has prohibited the transfer of enrichment or reprocessing equipment altogether. These are examples of how both nuclear suppliers achieve similarly strong protections against nuclear proliferation, albeit through different means.

In terms of safeguards as a prerequisite of supply, however, there is some discrepancy between the two nuclear powers. Of the 123 Agreements signed by the United States since 1997, 85 percent require the IAEA's Additional Protocol (AP) to Comprehensive Safeguards Agreements, whereas only 38 percent of Russian cooperative agreements over the same period include language requiring the AP.[31] In one recent example, Russia agreed to supply Egypt for its construction of the El-Dabaa nuclear power plant in 2015 without requiring Egypt to sign the AP. In another example, Russia has not encouraged Belarus to sign the AP, despite providing substantial supplies and guidance in the ongoing construction of the country's first nuclear power plant, in Ostrovets. The AP was initially created as a voluntary measure and has historically been controversial among non-nuclear weapon states due to the tit-for-tat nature of enhanced safeguards in exchange for nuclear weapons states' disarmament. However, perhaps the time has come to adjust this thinking and to incorporate the AP in agreements as a prerequisite to the supply of peaceful nuclear technology and material.

---

30   "NuScale Power Signs Agreement with Nuclearelectrica and Owner of Preferred Site for First SMR Site in Romania," NuScale, Press Release, May 23, 2022, https://newsroom.nuscalepower.com/press-releases/news-details/2022/NuScale-Power-Signs-Agreement-with-Nuclearelectrica-and-Owner-of-Preferred-Site-for-First-SMR-Site-in-Romania/default.aspx.

31   Stulberg and Darsey, "Moving Beyond Self-Restraint," 91.

The successful deployment of SMRs for carbon-free power generation around the world will depend on states' adherence to effective safeguards. In fact, safeguards must be considered even during the initial design phase of any new advanced nuclear technology, lest the design pose a challenge to safeguards inspectors when it comes to nuclear fuel accounting or accessibility once the reactor is operational. Nuclear suppliers must therefore ensure that new technology designs and deployment work in concert with the increased emphasis on safeguards in supply agreements, including AP language.

The inclusion of AP language is not the only area for improvement in nuclear supply agreements. None of the agreements provided by any nuclear supplier today require partner countries to be a signatory to any nuclear treaty or convention beyond the NPT, such as the Convention on the Physical Protection of Nuclear Material, the International Convention for the Suppression of Acts of Nuclear Terrorism, or the Convention on Nuclear Safety. Requiring that recipients of nuclear material or technology adhere to these additional frameworks would strengthen the overall security guarantees embedded in supply agreements. This should be standardized across the civil nuclear industry as the use of nuclear power increases and the associated material and equipment becomes more accessible to more countries, particularly through the use of SMRs.

Furthermore, obtaining nuclear fuel and reactor technology alone does not automatically incorporate nuclear power into a country's energy mix. States must also devote significant resources to establishing the regulatory and security infrastructure to support nuclear power generation and to effectively integrate it into the energy grid. The IAEA already provides some guidance and training for states that wish to develop this framework through Advisory Service Missions, but nuclear newcomers would be further supported if their nuclear cooperation agreements included the same from their suppliers. The United States and Russia have the world's most extensive nuclear research and development infrastructure, acquired and expanded over time through trial and error across multiple generations of nuclear technology. SMR designs from U.S.-based companies such as NuScale and Russia's state-controlled Rosatom are already being implemented in real-world settings and are likely to be the first available for export. No state is better equipped to serve as a model and provider of best practices to nuclear newcomers than the United States or Russia—and, as two of the world's largest carbon emitters, both states should be responsible for promoting clean energy in the face of climate change. As SMR technology continues to develop and availability increases along with global demand, supply agreements should provide training and institutional support for the corresponding infrastructure, in addition to the equipment itself. This is particularly necessary in cases where the recipient is new to the nuclear power "club" and is developing security and regulatory processes from scratch. Some of the first forays into SMR applications, such as NuScale's CFPP, will provide valuable lessons and precedents for future adopters of similar technology.

As mentioned earlier in this paper, it is worth noting that Russia's 2022 invasion of Ukraine holds implications for the nuclear industry from a security standpoint as well. The takeover of the Chernobyl nuclear site by Russian troops, where radiation levels have been reported to have recently increased, and Russian usurpation and military shelling of and around the Zaporizhzhia nuclear power plant—the largest nuclear plant in Europe—serve as reminders of the need for both secure nuclear facilities and effective security practices and procedures.[32] In the event that a country newly

---

32   Vasco Cotovio, Frederik Pleitgen, Byron Blunt, and Darya Markina, "Ukrainians Shocked by 'Crazy' Scene at Chernobyl After Russian Pullout Reveals Radioactive Contamination," CNN, April 9, 2022, https://www.cnn.com/2022/04/08/europe/chernobyl-russian-withdrawal-intl-cmd/index.html; Ivan Nechepurenko, "Russia Confirms that it Controls the Former Chernobyl Nuclear Plant," *New York Times*, February 25, 2022, https://www.nytimes.com/2022/02/25/world/europe/cher-

in possession of nuclear power should experience a national security emergency—be it a military invasion or a natural disaster brought on by climate change—it is imperative that the security of its nuclear material be assured. As of this writing, the IAEA is currently attempting to establish on-site inspections and monitoring of the facilities at Zaporizhzhia to prevent a nuclear disaster caused by the surrounding conflict. Having standards for the physical protection of nuclear material in place since the original supply agreement, supported by ongoing regulatory oversight which has been trained by a well-established and legally invested supplier-adviser, is the best way to ensure stability in times of crisis.

To achieve this, supply agreements should draw from external, parallel channels of cooperation in the nuclear sphere. NCMOUs are an excellent way to initiate and build relationships between countries' various stakeholders so that the regulatory, legal, scientific, and security infrastructure can be built out and supported before any kind of nuclear application or actual supply agreement takes place (this is part of the goal of the Biden administration's FIRST program). Establishing relationships in government or scientific communities early in the development process of SMRs and other advanced nuclear technology benefits both the eventual recipient and the producer of the equipment and makes it easier to incorporate safety, security, and nonproliferation measures through every stage of the process. As SMR production continues to grow in countries such as the United States, supported by government investment as well as market demand, perhaps a future "SMR Summit" may enable scientific and regulatory communities to exchange ideas on advanced safety measures and secure implementation of their designs. Additionally, existing multilateral fora such as the P5 Process (an informal convening of the NPT's recognized nuclear weapons states: the United States, the United Kingdom, France, Russia, and China) or the NSG should encourage and facilitate information sharing when it comes to best practices so that countries interested in someday joining the nuclear power club are already informed on what is expected for safe and secure implementation.

CONCLUSION

Given the rate of climate change and the increasing threat of its effects, the world needs an effective solution soon. Nuclear power offers such a solution through its reliable, efficient energy production with zero carbon emissions. The only way to combat climate change at the scale necessary to prevent its worst possible outcomes is to integrate more nuclear energy into energy grids around the world, thus reducing dependence on fossil fuels without limiting the amount of energy available to the world's population. SMRs, utilizing both existing and advanced nuclear technology, can make the switch to nuclear easier and more affordable for countries new to the nuclear power club. If the early applications of SMR plants currently underway, such as NuScale's CFPP, are successful, they will set a precedent for the rapid expansion of the use of such technology around the world. Therefore, this article has assumed that the advent of such technology will be followed by its more widespread adoption.

However, the spread of peaceful nuclear technology must not result in the spread of nuclear weapons programs or nuclear accidents. It is therefore critical that nuclear supply agreements keep up with increased reliance on nuclear power by incorporating stronger nonproliferation and security measures, such as language supportive of the IAEA's Additional Protocol, obligatory membership in other security treaties, and enhanced support for emerging nuclear security and regulatory frameworks.

nobyl-nuclear-plant-control.html; and AP, "Europe's Largest Nuclear Power Plant on Fire After Russian Shelling," *Politico*, March 3, 2022, https://www.politico.com/news/2022/03/03/europes-nuclear-power-plant-fire-russian-shelling-00014080.

This paper does not seek to imply that all the obstacles to combatting climate change can be resolved with the use of advanced nuclear technology alone. There are socioeconomic and political dimensions to both climate change and the spread of nuclear power which vary between countries. As mentioned earlier, there are still debates over effective strategies for the long-term storage of spent nuclear fuel. The current context of Russia's invasion of Ukraine also poses a new challenge related to the supply of high-assay, low-enriched uranium (HALEU) fuel, which is required for most advanced reactor designs and is currently supplied almost exclusively by Russia. It will take time for other countries, such as the United States, to develop their own supply of HALEU fuel, and this will add another layer of complexity to supply agreements. These and other developing issues must be resolved as the nuclear industry continues to expand. However, valuable strides are being made, and the potential rewards make tackling such tough questions worthwhile.

The growth of the nuclear power industry around the world has the potential to yield immense benefits for the world's population and economies. Nuclear weapons have been viewed since the end of World War II as the world's doomsday weapons: capable of annihilating millions of lives and posing existential danger to humanity. The prevention of nuclear holocaust must continue through arms control agreements and extending the current nuclear peace, but the use of nuclear technology for clean energy production should also be pursued as a means for mitigating climate change. With adjustments to bolster nuclear supply agreements, and continued government support for private investment in advanced nuclear technology such as SMRs, the world may come closer to achieving both priorities.

The Hypersonic Weapon Threat

Resilient Deterrence despite Destabilizing Buzzwords

By Wil Powell[1]

INTRODUCTION

Long theorized but technologically exhaustive to develop, hypersonic weapons have sensationalized the modern context of deterrence. In an era of great power competition, China, Russia, and the United States are each chasing newly realized advancements in hypersonic technologies to achieve their strategic objectives. Reports indicate that China and Russia have successfully fielded operational hypersonic weapons, while the United States is not likely to field an operational capability of their own until sometime in 2023.[2] Furthermore, it is well accepted that there are currently no air defenses with substantial capability to intercept a hypersonic weapon.[3] The capabilities provided by this next generation of weapon offer competitive advantages fleetingly achieved throughout history, but they should be recognized as nothing more than a sensationalized gimmick when leveraged against the global nuclear landscape.

A renewed arms race, centered on the ability of hypersonic weapons to provide time-sensitive nuclear strike deep into enemy territory, has once again reached headlines around the world. Russia has already demonstrated combat deployment of hypersonic weapons during their invasion of Ukraine.[4] This represents a disruptive capability in the hands of a state battered by economic

---

1. Wil Powell is a bomber instructor pilot in the U.S. Air Force. He is currently charged with training initial bomber crews with the knowledge and skills required to execute conventional and nuclear warfare. The views presented are his own and do not represent the official policy or position of the U.S. Department of Defense or its components.

2. Kelley M. Sayler, *Hypersonic Weapons: Background and Issues for Congress*, CRS Report No. R45811 (Washington, DC: Congressional Research Service, March 2022), https://sgp.fas.org/crs/weapons/R45811.pdf.

3. Ibid.

4. "Russia Claims First Use of Hypersonic Kinzhal Missile in Ukraine," BBC News, March 19, 2022, https://www.bbc.com/news/world-europe-60806151; and Tara Copp, "Russia Has Fired between '10 and 12' Hypersonics into Ukraine, Pentagon Says,"

sanctions and increasing diplomatic isolation. China has continued proliferation of nuclear warheads despite stating its intent to keep "its nuclear capabilities at the minimum level required for national security."[5]

According to Admiral Charles Richard, commander of U.S. Strategic Command (STRATCOM), China "likely intends to have at least 1,000 warheads by 2030, greatly exceeding previous DoD estimates."[6] Continued production by China—unbound by non-proliferation treaties—could spur a contested strategic security environment between three nuclear peer states. The ominous threat of nuclear-tipped hypersonic first strike yields new strategic stability implications. Do hypersonic weapons pose an existential threat to deterrence? This new capability, while a conventional novelty, will not drive a deep change in deterrence. In order to better understand the resiliency of deterrence, it is paramount to view the issue through the threat's capabilities, history, motivations, and probable solutions.

BACKGROUND

In a broad sense, hypersonic weapons may be understood as a category of weapon including two key characteristics: hypersonic velocity, which is defined as a speed above Mach 5, and cruise-missile-like maneuverability.[7] Conservatively, hypersonic speed reduces the amount of time an adversary has to detect and react to an incoming missile by 84 percent, out-racing most modern surface-to-air missiles (SAMs).[8] Unlike with ballistic missiles, which also travel at hypersonic velocities, the maneuverability of hypersonic weapons further enables them to bypass enemy missile defenses and complicates missile targeting.[9] While a ballistic missile travels at hypersonic velocity, its exo-atmospheric altitude and predictable trajectory allow for more simplified targeting compared to the maneuverability and endo-atmospheric flight of hypersonic weapons.[10]

In 2018, Michael Griffin, former undersecretary of defense for research and development, noted that "hypersonic targets are 10 to 20 times dimmer than what the U.S. normally tracks by satellites in geostationary orbit."[11] With current space-based tracking layers unable to effectively track a hypersonic weapon, the point of earliest detection may be within line of sight of a terrestrial radar station. Compared to a ballistic missile, hypersonic weapons can get significantly closer to a target prior to detection.[12]

Defense One, May 10, 2022, https://www.defenseone.com/threats /2022/05/russia-has-fired-between-10-and-12-hypersonics-ukraine-pentagon-says/366748/.

5   PRC State Council Information Office, *China's National Defense in the New Era* (Beijing: PRC State Council Information Office, July 2019), https://www.airuniversity.af.edu/Portals/10/CASI/documents/Translations/2019-07%20PRC%20White%20Paper%20on%20National%20Defense%20in%20the%20New%20Era.pdf?ver=akpbGkO5ogbDPPbflQkb5A%3D%3D; and Caitlin Campbell, *China's Military: The People's Liberation Army (PLA)*, CRS Report No. R46808 (Washington, DC: Congressional Research Service, June 2021), https://crsreports.congress.gov/product/pdf/R/R46808.

6   U.S. Congress, House, "Fiscal Year 2023 Strategic Forces Posture Hearing," Hearing before the House Armed Services Committee, 117th Cong., 2nd sess., March 1, 2022, https://armedservices.house.gov/2022/3/subcommittee-on-strategic-forces-hearing-fiscal-year-2023-strategic-forces-posture-hearing.

7   Susan Davis, "Hypersonic Weapons - A Technological Challenge for Allied Nations and NATO?," Science and Technology Committee, NATO Parliamentary Assembly, June 18, 2020, https://www.nato-pa.int/download-file?filename=sites/default/files/2020-07/039%20STC%2020%20E%20-%20HYPERSONIC%20WEAPONS.pdf.

8   Ibid. Data based on a traditional cruise missile at 0.8 mach compared to an HCM at the minimum qualifier, 5 mach.

9   Ibid.

10   Tom Karako and Masao Dahlgren, *Complex Air Defense: Countering the Hypersonic Missile Threat* (Lanham, MD: Rowman & Littlefield, February 2022), https://www.csis.org/analysis/complex-air-defense-countering-hypersonic-missile-threat.

11   U.S. Congress, Senate Committee on Armed Services, "Testimony of Michael Griffin," Hearing on New Technologies to Meet Emerging Threats, April 18, 2018, https://www.armed-services.senate.gov/imo/media/doc/18- 40_04-18-18.pdf.

12   For a thorough analysis of the threats hypersonic weapons pose as well as defense methods available and proposed to defeat them, please see Karako and Dahlgren, *Complex Air Defense*.

There are two primary types of hypersonic weapons: hypersonic glide vehicles (HGVs) and hypersonic cruise missiles (HCMs). HGVs may be surface- or air-launched and utilize a rocket booster to climb to the edge of earth's atmosphere before separating and gliding to their target.[13] An HCM differs in its use of a high-speed, air-breathing engine—known as a supersonic combustion ramjet, or scramjet—enabling powered hypersonic flight to a target.[14] Current cruise missiles operate at a much lower cruising altitudes and airspeeds than HCMs.[15] Subsonic (less than Mach 1) cruise missiles can also leverage stealth to their advantage due to a reduced infrared signature and greater flexibility in aerodynamic design.[16] Hypersonic missiles are at a disadvantage in this regard, though their advanced velocity and reduced warning time weigh in their favor for weapon survivability.

Hypersonic technology is not a new development, but recent efforts by many nations to research, test, and develop these weapons have grown significantly. The United States, Russia, and China are among the three states with the best-funded and most advanced programs. Other nations conducting research include India, Australia, Japan, and France as well as several additional European countries.[17] Hypersonic weapons pose a significant threat to countries that already possess a robust integrated air defense system, which means their development is viewed as a provocative innovation by all great power nations.

DETERRENCE

Nuclear deterrence, broadly speaking, relies on effectively coercing an adversary not to attack through the credible threat of an equally destructive response; in other words, an attack by an adversary would be met with overwhelming and punishing retaliation.[18] Reinforcing this notion, the Obama administration formally stated, "The United States will continue to ensure that … the perceived gains of attacking the United States or its allies and partners would be far outweighed by the unacceptable costs of the response."[19] Escalating tensions and, most visibly, U.S. constant airborne nuclear alert during the Cold War put this prominent strategy on display in an exceptionally public manner. In a nuclear conflict, deterrence by punishment is backed by a response to nuclear attack that is exponentially worse than the benefits of a successful attack.

In a message to Congress shortly after the end of World War II, President Harry S. Truman stated, "Never in history has society been confronted with a power so full of potential danger and at the same time so full of promise for the future of man and for the peace of the world."[20] The foundation of nuclear deterrence was set in these remarks. To this point, the leaders of the five nuclear weapon states, as recognized in the Treaty on the Non-Proliferation of Nuclear Weapons, issued a joint statement saying, "nuclear war cannot be won and must never be fought … nuclear weapons—for as

---

13   Richard H. Speier, George Nacouzi, Carrie Lee, and Richard M. Moore, *Hypersonic Missile Nonproliferation: Hindering the Spread of a New Class of Weapons* (Santa Monica, CA: RAND Corporation, 2017), https://www.rand.org/pubs/research_reports/RR2137.html.

14   Ibid., xi–xii.

15   Ibid.

16   Karako and Dahlgren, *Complex Air Defense*, 13–14.

17   Speier, Nacouzi, Lee, and Moore, *Hypersonic Missile Nonproliferation*, xii.

18   "Deterrence 101 Module 1 - Foundations of Deterrence," YouTube video, posted by CSIS, 28:44, https://www.youtube.com/watch?v=g1th_3vlLd4.

19   U.S. Department of Defense, *Nuclear Posture Review* (Washington, DC: April 2010), https://dod.defense.gov/Portals/1/features/defenseReviews/NPR/2010_Nuclear_Posture_Review_Report.pdf.

20   Harry S. Truman, "Message to Congress on the Atomic Bomb," atomicarchive.com, October 3, 1945, https://www.atomicarchive.com/resources/documents/deterrence/truman.html.

long as they continue to exist—should serve defensive purposes, deter aggression, and prevent war."[21] Treaties and other less-binding agreements have been successful in curbing incentives to use nuclear weapons. In fact, such agreements are actively used to extend nuclear deterrence from nuclear weapon states to other allied nations.[22] In the 78 years since its inception, nuclear deterrence has repeatedly proven its resilience.

Failure of deterrence would lead to a disruption in the national security of all nations involved. The previous two commanders of STRATCOM have emphasized that if deterrence were to fail, the top priority is reestablishing strategic stability between nations.[23] Strategic stability may be thought of as an equilibrium in the scales of deterrence where each state is mutually vulnerable to the capabilities of the other. The top priority, however, is to ensure deterrence does not fail.

The thinking goes that if a nuclear weapon state perceives a first strike as advantageous, there is little to deter them. Some believe options for nonproliferation, such as treaties or unilateral export controls, are the best route toward reducing crisis instability.[24] Establishing predictable international norms and implementing limitations on weapon quantities and utilization are among the strongest agreed-upon facets of crisis stability among scholars and public officials. If trust between great power nations continues deteriorating, however, it begs the question of what deterrence might look like in a state of imbalance or crisis.

HISTORICAL ADVANTAGES IN THE NUCLEAR AGE

While the capabilities of hypersonic weapons can negate enemy air defenses, thus neutralizing the ability to defend against a nuclear first strike, history reveals similar imbalances in the scales of deterrence. During World War II, bombers often faced heavy resistance from curtains of anti-aircraft artillery. This required a tremendous amount of artillery and still appeared to serve as only a minor deterrent to bombing missions during the war.[25] The superiority of a robust air-raid offensive necessitated high-altitude bombers by the hundreds to dismantle the adversary's ground-based artillery deterrent. With the realization of nuclear weapons shortly following the war, nations raced on two fronts: one to develop long-range, nuclear-capable jet bombers, and the other to develop systems aimed at efficiently destroying these bombers. Thus, there was a race to create SAMs to eliminate the competitive advantage adversaries possessed with nuclear-capable bombers.

These rapid technological advancements spurred an arms race for a superior nuclear arsenal. In rapid succession, the United States and Russia grew their bomber arsenals as well as their nuclear stockpiles in an effort to gain advantage over the other.[26] Eventually, a period of parity ensued between the two nations where neither was assured a winning advantage due to the number of thermonuclear devices developed and the number of delivery platforms, which were capable of

21 "Joint Statement of the Leaders of the Five Nuclear-Weapon States on Preventing Nuclear War and Avoiding Arms Races," The White House, January 3, 2022, https://www.whitehouse.gov/briefing-room/statements-releases/2022/01/03/p5-statement-on-preventing-nuclear-war-and-avoiding-arms-races/.

22 Paul Huth and Bruce Russett, "Deterrence Failure and Crisis Escalation," *International Studies Quarterly* 32, no. 1 (1988): 29–45, doi:10.2307/2600411.

23 Richard, "Statement before House Armed Services Committee"; and John E. Hyten, Statement Before the Senate Committee On Armed Services, February 26, 2019, https://www.armed-services.senate.gov/imo/media/doc/Hyten_02-26-19.pdf.

24 Speier, Nacouzi, Lee, and Moore, *Hypersonic Missile Nonproliferation*, xiii.

25 N.W. Routledge, *Anti-Aircraft Artillery*, 1914-55 (London: Brassey's, 1994).

26 "Report to the President's Foreign Intelligence Advisory Board (FIAB) on Intelligence Activities Relating to the Cuban Missile Build-up 'The Killian Report,'" JFK Library, March 28, 1962–January 7, 1963, https://www.jfklibrary.org/asset-viewer/archives/JFKNSF/471/JFKNSF-471-005.

overwhelming either country's defenses. Not long after, just as the "Killian Report" hypothesized, a new weapon system was realized, delivering a renewed position of advantage to the U.S. military.[27] Engineers discovered a way to boost a nuclear weapon to space, separate the kinetic package, and deliver it ballistically to earth with striking accuracy. The newly developed intercontinental ballistic missile (ICBM) once again allowed delivery of nuclear payloads and denied an ability to deter. The response spurred multiple advancements in ballistic missile defense (BMD) technologies during the Cold War to shorten the gap in offensive and defensive capabilities.

As it became clear that the cycle of advancing counter-offensive and defensive technologies would continue, the Anti-Ballistic Missile (ABM) Treaty was adopted between the United States and Soviet Union in 1972 to stymie the ongoing arms race. This agreement remained in effect until the United States withdrew from the treaty in 2001, citing the threat of terrorism or rogue states.[28] Many foreign and domestic critics expressed concern over this move, saying increasing current BMD inventories would cause fears of a U.S. nuclear first strike.[29]

Once again, historians and experts are questioning the broadening gap in military ability with the proliferation of hypersonic weapons. A weapon with no known defense may entice a subsequent arms race and stockpiling. A nation with a sizeable advantage in this arena is likely to be searching for the ideal time to execute a decapitating first strike against an adversary. In fact, the lack of defenses, reduced warning and reaction timeline, inexact ability to determine enemy weapon type, and inability to determine a designated target could force a more responsive defensive posture. Delegation of command and control may result in a "launch on warning" order, like that of the Cold War, which brought the world to the brink of catastrophe.

GREAT POWER HYPERSONIC PROGRAMS

RUSSIA

Of the three great power nations actively testing hypersonic missiles, Russian state media and diplomats continually tout their program as the world's most advanced. In Admiral Richard's March 2022 statement before the House Armed Services Committee, he states Russia's "Avangard HGV, Tsirkon hypersonic anti-ship and land-attack missile, and Kinzahl ALBM (air-launched ballistic missile) are operationally fielded."[30] Within weeks, Russia supported these claims with the first combat use of the Kinzahl missile on a Ukrainian arms depot.[31] It should be noted, however, that the Kinzahl is a primitive hypersonic weapon lacking maneuverability, and there is debate as to whether it meets the criteria for inclusion in the category altogether.[32] However, it is of note due to its nuclear capability and its ability to be air-launched and to travel at hypersonic velocity.

27   Ibid.

28   "U.S. Withdrawal From the ABM Treaty: President Bush's Remarks and U.S. Diplomatic Notes," Arms Control Association, January 2002, https://www.armscontrol.org/act/2002-01/us-withdrawal-abm-treaty-president-bush%E2%80%99s-re-marks-us-diplomatic-notes.

29   William James Perry, *My Journey at the Nuclear Brink* (Stanford, CA: Stanford University Press, 2015).

30   Charles Richard, "STRATCOM Posture Statement," Statement of Charles A. Richard Commander United States Strategic Command Before the House Armed Services Committee On Strategic Forces, March 1, 2022, https://www.stratcom.mil/Portals/8/Documents/2022%20USSTRATCOM%20Posture%20Statement.pdf?ver=CUIoOCLyos9xe9C9I0XjMQ%3D%3D.

31   "Russia Claims First Use of Hypersonic Kinzhal Missile in Ukraine," BBC News, March 19, 2022, https://www.bbc.com/news/world-europe-60806151.

32   Alex Hollings, "Why the 'Hypersonic Missile' Russia Says It Just Used in Ukraine Isn't as Advanced as It Sounds," Business Insider, March 21, 2022, https://www.businessinsider.com/russias-kinzhal-hypersonic-missile-isnt-as-advanced-as-it-sounds-2022-3.

Russia's Avangard was reportedly first operational in 2018 after 14 flight tests spanning 28 years.[33] It is reported to be a boost-glide HGV carried to an apogee of 100 kilometers by the silo-based SS-19 Stiletto before release. Future developments will have the missile carried by the R-28 Sarmat booster.[34] The estimated range of the weapon is 6,000 kilometers, making it an ICBM-class weapon. At such a range, the North American mainland remains out of reach, but the entirety of Europe could be struck within minutes.

Additional hypersonic developments include the 3M22 Tsirkon, known by its North Atlantic Treaty Organization's (NATO) reporting as the SS-N-33, an anti-ship, land-attack HCM. It has seen successful ship-based launches off the *Admiral Gorshkov* as well as from the nuclear-powered *Severodvinsk*, a Yasen-class submarine, in both surface and submerged tests.[35] The Tsirkon will be capable of traveling at Mach 9 and striking targets up to 1,000 kilometers away.[36]

Russia often cites U.S. withdrawal from the Anti-Ballistic Missile (ABM) Treaty and mobilization of missile defenses both domestically and worldwide as motivating factors for accelerating their weapons development.[37] Underpinning these motivations, President Vladimir Putin stated in 2018 that "Russia was categorically against (U.S. withdrawal from the ABM Treaty). We saw the Soviet-U.S. ABM Treaty signed in 1972 as the cornerstone of the international security system."[38] To further his case, Putin went on to say the United States "is permitting constant, uncontrolled growth of the number of anti-ballistic missiles, improving their quality, and creating new missile launching areas. If we do not do something, eventually this will result in the complete devaluation of Russia's nuclear potential."[39] He went on to announce the Sarmat missile, advocating its hypersonic velocity, maneuverability, and nuclear capability.

CHINA

Chinese national defense policy states that their people, as a peace-loving nation, have suffered from a history of aggression and war and further states that "China will never inflict such sufferings on any other country."[40] Accusations of bellicose U.S. encroachment on Chinese sovereignty in the Indo-Pacific has led to China declaring a need to develop hypersonic weapons. In particular, China's pursuit of hypersonic weapons reflects a fear of the U.S. ability to conduct a disarming preemptive strike on China's comparatively smaller nuclear arsenal. China fears a subsequent retaliatory strike would achieve limited success given U.S. missile defenses in the region.[41] These fears have led China to

---

33   Pavel Podvig, "Avangard system is tested, said to be fully ready for deployment," Russian Forces, December 26, 2018, http://russianforces.org/blog/2018/12/avangard_system_is_tested_said.shtml.

34   Nikolay Surkov, "Hypersonic Avangard," *Izvestia*, March 2, 2018, https://iz.ru/715170/nikolai-surkov/giperzvukovoi-avangard; and Missile Defense Project, "Avangard," Missile Threat, CSIS, January 3, 2019, last modified July 31, 2021, https://missilethreat.csis.org/missile/avangard/.

35   "Russia test-launches Tsirkon hypersonic missile from ship for first time," TASS, February 27, 2020, https://tass.com/defense/1124339; and "Tsirkon 'Serial Production': Russia's Race for Hypersonic Missiles Revealed," WION, February 10, 2022, https://www.wionews.com/photos/tsirkon-serial-production-russias-race-for-hypersonic-missiles-revealed-452075#putins-invincible-weapons-452052.

36   "Tsirkon 'Serial Production'," WION.

37   Sayler, Hypersonic Weapons, 12.

38   Vladimir Putin, "Presidential Address to the Federal Assembly," Kremlin, March 1, 2018, http://en.kremlin.ru/events/president/news/56957.

39   Ibid.

40   PRC State Council Information Office, *China's National Defense in the New Era*, 7–8.

41   Tong Zhao, "Conventional Challenges to Strategic Stability: Chinese Perception of Hypersonic Technology and the Security Dilemma," Carnegie Endowment, July 23, 2018, https://carnegieendowment.org/2018/07/23/conventional-challenges-to-strategic-stability-chinese-perceptions-of-hypersonic-technology-and-security-dilemma-pub-76894; and Lora Saalman, "China's Calculus on Hypersonic Glide," Stockholm International Peace Research Institute, August 15, 2017, https://www.sipri.org/commentary/topical-backgrounder/2017/chinas-calculus-hypersonic-glide.

build up its anti-access/area denial (A2/AD) capabilities, in part to protect regional territory but also as a means to stymie perceived U.S. aggression.

Efforts to advance hypersonic capabilities have caught several officials off guard.[42] As recently as August 2021, China demonstrated the ability to place an HGV in orbit using their Long March ICBM-class rocket.[43] The Long March was able to orbit the earth before releasing an HGV toward a target, a process known as fractional orbital bombardment (FOB).[44] Such a capability has garnered attention from U.S. senior leadership. In March 2022, Admiral Richard stated "The PRC's pursuit of an ICBM delivered HGV with FOB capability is a technological achievement with serious implications for strategic stability."[45] Such a capability appears counter to one of "safeguarding national territorial sovereignty and maritime rights and interests," as stated in Chinese policy.[46]

A regional DF-17 ballistic missile, designed to launch the DF-ZF HGV, is assessed by U.S. defense officials to have achieved operational status in 2019.[47] This system could strike targets within 1,000 to 1,500 miles, providing a credible deterrent to all regional opposition.[48] In addition to the DF-17, the DF-41, a launch vehicle providing ICBM range, has been undergoing modification to carry HGVs capable of conventional or nuclear payloads.[49] Additional reports were made in August 2018 that China had conducted a test flight of a hypersonic "wave rider" aircraft, Starry Sky-2.[50] Such a vehicle could penetrate deep into enemy defenses and deliver multiple payloads, similar to FOB but at a much lower altitude.

UNITED STATES

Policymakers have been consistent in their commitment that U.S. hypersonic weapons will only be conventionally armed, contrary to the nuclear capabilities of the Russian and Chinese programs.[51] Such a commitment will greatly reduce the risk calculus associated with an adversary inadvertently categorizing a U.S. hypersonic launch as a weapon of mass destruction, though nuclear-tipped missiles can still be militarily effective with 100-times less accuracy than that of a conventional missile.[52] The problem, therefore, is creating a hypersonic weapon capable of delivering a payload within a few meters of its designated target. Nevertheless, U.S. officials are researching a multitude of weapon designs, with the intent to deter adversaries and prevent their ability to use A2/AD.[53]

No official program of record is said to exist for U.S. hypersonic weapons, but several such systems are in research, development, test, and evaluation (RDT&E) for an operational prototype.[54] The U.S.

42 Demetri Sevastopulo and Kathrin Hille, "China Tests New Space Capability with Hypersonic Missile," *Financial Times*, October 16, 2021, https://www.ft.com/content/ba0a3cde-719b-4040-93cb-a486e1f843fb.

43 Ibid.

44 Ibid.

45 Richard, "STRATCOM Posture Statement," 6.

46 PRC State Council Information Office, *China's National Defense In a New Era*, 11.

47 Sayler, Hypersonic Weapons, 17.

48 Ibid.

49 U.S.-China Economic and Security Review Commission, *2018 Annual Report to Congress* (Washington, DC: 2018), 235, https://www.uscc.gov/sites/ default/files/annual_reports/2018%20Annual%20Report%20to%20Congress.pdf.

50 Liu Zhen, "China's Hypersonic Aircraft Can Fire Nukes at 6 Times the Speed of Sound," *South China Morning Post*, August 6, 2018, https://www.scmp.com/news/china/diplomacy-defence/article/2158524/chinas-hypersonic-aircraft-starry-sky-2-could-be-used.

51 Sayler, Hypersonic Weapons, 4.

52 James M. Acton, "China's Advanced Weapons," Testimony to the U.S. China Economic and Security Review Commission, February 23, 2017, https://carnegieendowment.org/2017/02/23/china-s-advanced-weapons-pub-68095.

53 House Armed Services Committee, *Future of Defense Task Force Report 2020* (Washington, DC: House Armed Services Committee, September 2020), https://armedservices.house.gov/_cache/files/2/6/26129500-d208-47ba-a9f7-25a8f-82828b0/424EB2008281A3C79BA8C7EA71890AE9.future-of-defense-task-force-report.pdf.

54 Steve Trimble, "New Long-Term Pentagon Plan Boosts Hypersonics, But Only Prototypes," *Aviation Week*, March 15, 2019,

Navy has taken the helm developing a common glide vehicle to be used by multiple services.[55] Once in production, the common HGV is expected to be mated with a Navy boost system to create their Conventional Prompt Strike missile, which will also be synonymous with the Army's Long-Range Hypersonic Weapon.[56] Such a weapon will provide a ranged, time-sensitive response option from land or sea to strike high-value adversary targets or disrupt enemy A2/AD links. The Army has stated that the range will be greater than 2,775 kilometers, placing Moscow within reach from London, or Taiwan within rapid response from Guam, for reference.[57]

The Navy has also released RDT&E plans for a hypersonic, air-launched, anti-ship missile referred to as the Offensive Anti-Surface Warfare (OASuW) Weapon Increment II, more concisely referred to as the Hypersonic Air-Launched OASuW (HALO).[58] This new weapon is being developed with an emphasis on enabling naval operations in contested littoral waters and A2/AD environments.[59]

The U.S. Air Force also plans to use the common HGV technology on the AGM-183 Air-Launched Rapid Response Weapon (ARRW).[60] It is assessed to reach targets 1,000 miles away within 12 minutes.[61] ARRW was also anticipated to achieve early operational capability in fiscal year (FY) 2022, but it has been postponed to FY 2023 and sized down to a single missile from the 12 that were originally funded.[62] While funding for ARRW procurement has been reabsorbed by RDT&E, the Air Force has newly requested $200 million in RDT&E funding for the Hypersonic Attack Cruise Missile (HACM).[63] Initial designs for the HACM may not be completed until FY 2023 at the earliest.[64]

Additional research is being conducted by the Defense Advanced Research Projects Agency (DARPA) to support hypersonic objectives for the Navy, Army, and Air Force. The Tactical Boost Glide (TBG) is an HGV design undergoing testing as both an air- and sea-based launch option.[65] Similarly, in order to support Army objectives, the TBG is undergoing evaluation for a ground-based launch alternative known as Operational Fires.[66] The final program DARPA has been working on jointly with the Air Force is an HCM known as the Hypersonic Air-breathing Weapon Concept (HAWC). HAWC will place an emphasis on sustained hypersonic cruise with a smaller missile body, enabling multiple air platforms to carry and deploy such a weapon.[67] A successful test of the HAWC reportedly occurred in March 2022 but was not publicly reported at the time amid tensions in the Russia-Ukraine conflict.[68] In total, $7.2 billion has been requested in the FY 2023 budget to support long-range fires, to include

---

https://aviationweek.com/defense/new-long-term-pentagon-plan-boosts-hypersonics-only-prototypes.

55   Sayler, Hypersonic Weapons, 5.

56   Ibid.

57   Sydney J. Freedberg, Jr., "Army Discloses Hypersonic LRHW Range Of 1,725 Miles; Watch Out China," Breaking Defense, May 12, 2021, https://breakingdefense.com/2021/05/army-discloses-hypersonic-lrhw-range-of-1725-miles-watch-out-china/.

58   "Department of Defense Fiscal Year (FY) 2023 Budget Estimates," Research, Development, Test & Evaluation, Navy, April 2022, https://www.secnav.navy.mil/fmc/fmb/Documents/23pres/RDTEN_BA4_Book.pdf.

59   Ibid.

60   Thomas Newdick, "Air Force Says New Hypersonic Missile Will Hit Targets 1,000 Miles Away In Under 12 Minutes," The Drive, October 13, 2020, https://www.thedrive.com/the-war-zone/37045/air-force-says-new-hypersonic-missile-will-hit-targets-1000-miles-away-in-under-12-minutes.

61   Ibid.

62   Valerie Insinna, "Air Force Ditches Plans to Buy First Hypersonic ARRW Missile in FY23," Breaking Defense, March 29, 2022, https://breakingdefense.com/2022/03/air-force-ditches-plans-to-buy-first-hypersonic-arrw-missile-in-fy23/.

63   Sayler, Hypersonic Weapons, 9.

64   Ibid.

65   Ibid.

66   Ibid.

67   Ibid.

68   Oren Liebermann, "US Tested Hypersonic Missile in Mid-March but Kept It Quiet to Avoid Escalating Tensions with Russia," CNN, April 4, 2022, https://www.cnn.com/2022/04/04/politics/us-hypersonic-missile-test/index.html.

U.S. hypersonic RDT&E, up from $6.6 billion the year prior.[69] Comparatively, $225.5 million has been requested in FY 2023 for hypersonic defense programs, down from $247.9 million the year prior.[70]

IMPACT TO DETERRENCE

The impact hypersonic weapons have on strategic stability is contested at best. It is unlikely that the low-altitude maneuverability hypersonic weapons offer will create an irreconcilable advantage for any nation. In the event of a large-scale nuclear strike, no great power nation possesses a sufficiently robust missile defense system to defend against an overwhelming strike.[71] Thus, adding many nuclear-capable hypersonic weapons will not fundamentally impact the balances in strategic capabilities. The threat of an equally destructive retaliatory strike remains, independent to hypersonic proliferation.

While the strategic landscape is likely to see change in the margins of policy, first use of tactical hypersonic weapons poses an increased threat to regional conflicts. The compressed reaction time to identify a launch and assess the payload, nuclear or not, leaves little room for error. This is a key issue for any dual-capable weapons system. The compressed reaction time compounds the need to address this growing concern. The possibility for pre-emptive strikes leaves NATO's Integrated Air and Missile Defense and command and control exposed.

Current authority to employ nuclear weapons lies either with the president alone or with a circle of senior decisionmakers in each great power nation. If rapid response efforts are made against hypersonic nuclear weapons, the possibility exists to delegate that authority below the necessary level, further increasing crisis instability. Such tactics occurred during the Cold War and nearly led to nuclear launches during multiple tense exchanges, all ultimately averted. This historical lesson makes it unlikely for policymakers to return to a such a potentially destructive stance. Instead, advances in automation and artificial intelligence (AI) should be leveraged for earlier identification, enabling senior leaders to maximize time for contemplating a course of action. Additional preplanned responses can also be utilized, similar to the Single Integrated Operational Plan response options prepared during the Cold War. Early identification efforts, to include intelligence warning of an imminent launch, will ultimately provide the greatest amount of time for decisionmakers to leverage all options.

69 Lloyd J. Austin "The Department of Defense Releases the President's Fiscal Year 2023 Defense Budget," U.S. Department of Defense, March 28, 2022, https://www.defense.gov/News/Releases/Release/Article/2980014/the-department-of-defense-releases-the-presidents-fiscal-year-2023-defense-budg/.

70 Department of Defense, *Department of Defense Fiscal Year (FY) 2023 Budget Estimates, Missile Defense Agency Defense-Wide Justification Book Volume 2a of 5* (Washington, DC: April 2022), 631, https://comptroller.defense.gov/Portals/45/Documents/defbudget/ fy2023/budget_justification/pdfs/03_RDT_and_E/RDTE_Vol2_MDA_RDTE_PB23_Justification_Book.pdf; Office of the Under Secretary of Defense (Comptroller)/Chief Financial Officer, *Defense Budget Overview: United States Department of Defense Fiscal Year 2023 Budget Request* (Washington, DC: Department of Defense, April 2022), 2–16, https://comptroller.defense.gov/Portals/45/Documents/defbudget/FY2023/FY2023_Budget_Request_Overview_Book.pdf; Department of Defense, *Department of Defense Fiscal Year (FY) 2022 Budget Estimates, Missile Defense Agency Defense-Wide Justification Book Volume 2a of 5* (Washington, DC: Department of Defense, 2022), 569, https://comptroller.defense.gov/Portals/45/Documents/defbudget/fy2022/budget_justification/pdfs/03_RDT_and_E/RDTE_Vol2_MDA_RDTE_PB22_Justification_Book.pdf; and Office of the Under Secretary of Defense (Comptroller)/Chief Financial Officer, *Defense Budget Overview: United States Department of Defense Fiscal Year 2022 Budget Request* (Washington, DC: Office of the Under Secretary of Defense, May 2021), 3-2, https://comptroller.defense.gov/Portals/45/Documents/defbudget/FY2022/FY2022_Budget_Request_Overview_Book.pdf.

71 Office of the Secretary of Defense, *Missile Defense Review* (Washington, DC: Department of Defense, 2019), v, https://media.defense.gov /2019/Jan/17/2002080666/-1/-1/1/2019-MISSILE-DEFENSE-REVIEW.PDF.

CONCLUSION

The complexity of countering the threat from hypersonic weapons has been met with much disagreement and varying solutions. Preventative and defensive measures have both been proposed, including using improved AI systems to aid in identifying, categorizing, and even responding to incoming hypersonic threats. Some nations have increased funding for space-based detection and tracking layers, mid-course and terminal-phase missile interceptors, and improved integrated air defense communications as protective measures. In addition to these proposed solutions, pursuing an arms control agreement between all parties would aid in defining the ambiguous purpose and uses for hypersonic weapons, as well as developing crisis channels for de-escalation.

As seen with the initial proliferation of nuclear devices and subsequent ripples in offensive and defensive engineering, hypersonic weapons are the next iteration of human military prowess. Short-term advantages may be noted, particularly in regional conflict, but as with historical strategic imbalances, stability is likely to remain. The credibility and capability of a survivable, retaliatory, and punishing second strike remains with or without hypersonic weapons.

U.S. Nuclear Posture Reviews consistently reiterate statements of deterrence. In 2010, the Obama administration declared "any use of nuclear weapons will be met with a response that would be effective and overwhelming."[72] Following suit, the Trump administration released a three-pronged approach to tailored deterrence, culminating with: "any nuclear escalation will fail to achieve their objectives, and will instead result in unacceptable consequences for them."[73] The comments by both administrations, and many before them, defy party lines to support a common nuclear deterrent.

Consistent with the previous administration, the 2022 Nuclear Posture Review reiterates the number one priority of U.S. nuclear weapons to be deterring strategic attacks. Their conclusion draws that "nuclear weapons are required to deter not only nuclear attack, but also a narrow range of other high consequence, strategic-level attacks."[74] The National Defense Strategy refers to hypersonics as a complex escalation dynamic for strategic stability.[75] Despite hypersonic weapons representing a conspicuous new capability, decades of national leadership have made it clear this emerging technology will not fundamentally shift the underlying principles of nuclear deterrence.

---

72 "Statement by President Barack Obama on the Release of Nuclear Posture Review," The White House, April 6, 2010, 33, https://obamawhitehouse.archives.gov/the-press-office/statement-president-barack-obama-release-nuclear-posture-review.

73 Department of Defense, *Nuclear Posture Review* (Washington, DC: Department of Defense, 2018), https://dod.defense.gov/News/Special-Reports/0218_npr/.

74 Department of Defense, *2022 National Defense Strategy* (Washington, DC: Department of Defense, October 2022), https://media.defense.gov/2022/Oct/27/2003103845/-1/-1/1/2022-NATIONAL-DEFENSE-STRATEGY-NPR-MDR.PDF

75 Ibid., 6.

WMD Assassinations

Another Tool in the Autocratic Toolbox

By Erica Symonds[1]

INTRODUCTION

In 1978, while waiting for a bus in London, Bulgarian defector Georgi Markov was jabbed in the thigh with the tip of an umbrella that released a pellet containing ricin. Markov died a few days later.[2] This incident became notorious during the Cold War given the air of mystery surrounding the covert use of a biological weapon to assassinate a political dissident in a public place.

In more recent years, the world has seen repeated assassinations and assassination attempts using substances traditionally categorized as weapons of mass destruction (WMDs) to target individuals presenting an apparent threat to an authoritarian regime. Creating and administering the poisons for these attacks requires skilled scientists, precise equipment, and technical know-how, not to mention funding and dedication, making them distinctly different from more traditional means of assassination. Like the umbrella attack, these cases gained worldwide attention. They simultaneously sent a message to those speaking out against the regime while seemingly allowing for plausible deniability by the state perpetrating the attack, which could create its own narrative.

This paper examines state-level use of deadly substances traditionally considered problematic for their potential to cause mass casualties but employed in recent years for targeted attacks on individuals posing a political threat. The decision to use chemical, biological, or radiological weapons for assassinations, particularly to target nationals or defectors in a foreign country, has implications for norms of behavior. To consider the global impact of WMD assassinations on norms and taboos against use, this paper reviews WMD norms and taboos, considers four recent cases of

---

1   Erica Symonds is a foreign affairs specialist at the U.S. Department of Energy's National Nuclear Security Administration (DOE/NNSA) in the Office of Nuclear Verification. The views expressed in this paper are those of the author and are not an official policy or position of DOE, NNSA, or the U.S. government.

2   Compiled by Richard Nelsson, "The poison-tipped umbrella: the death of Georgi Markov in 1978 – archive," *The Guardian*, September 9, 2020, https://www.theguardian.com/world/from-the-archive-blog/2020/sep/09/georgi-markov-killed-poisoned-umbrella-london-1978.

WMD assassination or attempted assassination, evaluates why attackers might use WMDs and the complexity in doing so and analyzes how the international community has responded. The paper concludes with an assessment of the erosion of norms due to WMD assassinations and the role of democracies in holding violators accountable.

WMD NORMS AND TABOOS

Any use of a biological, chemical, or radiological weapon is problematic for violating international norms and taboos against the use of WMDs.[3] The Biological Weapons Convention (BWC) and Chemical Weapons Convention (CWC), which entered into force in 1975 and 1997, respectively, ban the development, production, acquisition, transfer, stockpile, and use of biological weapons (BWs) and chemical weapons (CWs). Both treaties have near universal participation by states parties. The CWC includes verification provisions requiring on-site inspections, while the BWC lacks a verification regime. The Organization for the Prohibition of Chemical Weapons (OPCW), the implementing body for the CWC, has a network of laboratories around the world that stand ready to analyze samples from alleged CW incidents.

Although a ban on the development or use of radiological weapons has been discussed repeatedly in UN forums, no such provision exists in an international treaty.[4] Radiological threats could range from an assassination of an individual or individuals to a small-scale blast from a radiological dispersal device or "dirty bomb." By contrast, nuclear weapons, have a much stronger destructive power and extensive radioactive fallout. A strong taboo against the use of nuclear weapons exists, and multilateral organizations and grassroots efforts continue to push nuclear powers to eliminate their nuclear weapons stockpiles.[5]

Some argue that the taboo against CW use has been "largely shattered" due to the range of CW attacks since 2012.[6] While norms against use have degraded, states continue working to hold violators accountable through international mechanisms and unilaterally. While BWs and radiological weapons continue to raise concerns, norms and the taboo against BW use and the taboo against radiological weapons use remain intact. WMD use in assassinations, however, creates daunting challenges for responsible governments through activities such as identification, medical response and treatment, decontamination to reduce risk to the public, and attribution, not to mention holding perpetrators accountable. As a whole, the international community is not tolerant of WMD use, yet effective accountability and consequences for use continue to pose challenges.

WMDs are by name associated with "mass destruction," and this term may seem out of place when discussing assassinations. This paper uses "WMD" to refer to a group of weapons that have the potential to cause mass destruction, but the substances that make up this type of weapon could also be used for targeted attacks on individuals. A distinguishing feature is that these substances are highly lethal and could cause a larger-scale crisis even when intended to target a particular

---

3   This paper uses "norms" and "taboos" as defined by: Rebecca K.C. Hersman and William Pittinos, *Restoring Restraint: Enforcing Accountability for Users of Chemical Weapons* (Lanham, MD: Rowman & Littlefield, June 2018), https://csis-website-prod. s3.amazonaws.com/s3fs-public/publication/180607_Hersman_RestoringRestraint_Web.pdf.

4   Daniil Kobyakov and Nicolas Florquin, "'Dirty Bomb' Threat Awakens Dormant Disarmament Conference," James Martin Center for Nonproliferation Studies at Middlebury Institute of International Studies at Monterey, August 26, 2002, https:// nonproliferation.org/dirty-bomb-threat-awakens-dormant-disarmament-conference-2/.

5   Nina Tannenwald, "Stigmatizing the Bomb: Origins of the Nuclear Taboo," *International Security* 29, no. 4 (Spring 2005): 5–49, https://www.jstor.org/stable/4137496.

6   Hersman and Pittinos, *Restoring Restraint*, 4–6.

individual. They can be difficult to handle and administer without inadvertently causing harmful public health effects, such as chemical or radiological contamination and denial of use of a specific location or larger geographic area.

Of the four cases this paper will highlight, one involved a radiological weapon and three involved CWs, two of which came from a CWC signatory and state party.

FOUR HIGHLY PUBLICIZED CASES

Since the mid-2000s, four poisonings using chemical or radiological weapons stand out for their notoriety (see Table 1), in large part due to the cross-border nature of the attacks, although numerous other assassinations and attempted assassinations have been reported or suspected in past decades involving biological, chemical, and radiological weapons. Notably, this is only a recent group of cases. States have long used poisons that would be considered BWs or CWs for assassinations to silence political dissidents, and terrorist organizations and rogue individuals have used WMD poisons as retribution against real or perceived opponents.[7]

Table 1: Overview of Highlighted Cases of WMD Assassinations or Attempted Assassinations

Target	Year	Location of Attack	Named State Perpetrator	Weapon Type	Suspected Reason	Outcome
Alexander Litvinenko	2006	United Kingdom	Russia	Radiological (polonium-210)	Regime critic	Died
Kim Jong-nam	2017	Malaysia	North Korea	Chemical (VX nerve agent)	Regime critic and potential leadership heir	Died
Sergei and Yulia Skripal	2019	United Kingdom	Russia	Chemical (Novichok nerve agent)	Regime critic	Survived
Alexei Navalny	2020	Russia (and treatment in Germany)	Russia	Chemical (Novichok nerve agent)	Regime critic	Survived

Source: Author's own research and analysis.

---

7   Stephen Burgess and Helen Purkit, *The Rollback of South Africa's Chemical and Biological Warfare Program* (Maxwell Air Force Base, AL: U.S. Air Force Counterproliferation Center, April 2001), https://www.airuniversity.af.edu/Portals/10/CSDS/Books/therollbackofsoafricachembio2.pdf; Glenn Cross, "Death in the Air: Revisiting the 2001 Anthrax Mailings and the Amerithrax Investigation," War on the Rocks, January 16, 2019, https://warontherocks.com/2019/01/death-in-the-air-revisiting-the-2001-anthrax-mailings-and-the-amerithrax-investigation/; Tim Ballard, Jason Pate, Gary Ackerman, Diana McCauley, and Sean Lawson, "Chronology of Aum Shinrikyo's CBW Activities," James Martin Center for Nonproliferation Studies at Middlebury Institute of International Studies at Monterey, August 26, 2005, https://nonproliferation.org/chronology-of-aum-shinrikyos-cbw-activities/; and Charles Streeper, Marcie Lombardi, and Lee Cantrell, "LA-UR 07-3686 Diversions of radioactive material and the difficulties in case tracking," Institute of Nuclear Materials Management, 2008, https://www.researchgate.net/publication/266386786_LA-UR_07-3686_Diversions_of_radioactive_material_and_the_difficulties_in_case_tracking.

POLONIUM-210: ALEXANDER LITVINENKO

In 2006, former Russian KGB officer Alexander Litvinenko drank green tea laced with polonium-210 at a London hotel and died three painful weeks later. Litvinenko was meeting with two Russian men for what he thought was a business meeting, but the men were in fact on a mission to assassinate him. While in the hospital, Litvinenko's story went viral, illustrated by a haunting photo of him in his hospital bed wearing a green hospital gown.[8] Russia likely wanted Litvinenko dead for accusing Russia's Federal Security Service, the KGB's successor, of being responsible for the 1999 apartment bombings that killed hundreds and which Russia attributed to Chechen separatists.[9] He was also trying to expose the Putin administration's links to organized crime and was reportedly working for the United Kingdom's foreign intelligence agency, MI6.[10]

The United Kingdom conducted a public inquiry which determined the series of events that occurred, identified the two men who killed Litvinenko, and pointed to a state-sponsored assassination by Russia.[11] The European Court of Human Rights (ECHR) found that Russia was responsible.[12] Litvinenko, who defected to the United Kingdom and became an outspoken critic of the Kremlin, believed Russian president Vladimir Putin directly ordered his death.[13] The ECHR found the use of polonium-210, a rare radioactive isotope, key because it "was an unlikely murder weapon for common criminals and must have come from a reactor under state control."[14]

The two Russian men who poisoned Litvinenko succeeded in administering a lethal dose on their third attempt and left a radioactive trail of their movements along the way, at times with alarming levels of contamination in public spaces. One journalist called them "idiotic, verging on suicidal" for their carelessness given how deadly polonium is once ingested and possibly indicating their lack of awareness of the nature of the poison.[15] After the second assassination attempt, one of the killers emptied a vial of polonium into the sink of his hotel bathroom, spilling it. The hotel room was like "a scene from an atomic horror story," with such high levels of radiation that the scientists wearing protective gear asked to leave the room after seeing the readings on their instruments.[16] The towels used to clean up the spill that detectives later found stuck in a laundry chute were so highly contaminated that they went to a UK nuclear facility for analysis. One of those towels was the most radioactive item discovered in the decade-long investigation by London police, containing enough radiation to kill an adult male within a month if absorbed into his blood.[17] The contamination from

8   Luke Harding, "Alexander Litvinenko: the man who solved his own murder," *The Guardian*, January 16, 2016, https://www.theguardian.com/world/2016/jan/19/alexander-litvinenko-the-man-who-solved-his-own-murder.

9   Robert Owen, *The Litvinenko Inquiry* (London: UK Home Office, January 2016), 54, 57–58, 239–44, https://assets.publishing.service.gov.uk/government/uploads/system/uploads/attachment_data/file/493860/The-Litvinenko-Inquiry-H-C-695-web.pdf.

10   Ibid., 212, 239.

11   Ibid., 227.

12   European Court of Human Rights, "Russia was responsible for assassination of Aleksandr Litvinenko in the UK," Press release, September 21, 2021, https://hudoc.echr.coe.int/app/conversion/pdf/?library=ECHR&id=003-7125276-9653243&filename=Judgment%20Carter%20v.%20Russia%20-%20Russia%20was%20responsible%20for%20assassination%20of%20Aleksandr%20Litvinenko%20in%20the%20UK.pdf.

13   Guy Faulconbridge and Michael Holden, "Russia was behind Litvinenko assassination, European court finds," Reuters, September 21, 2021, https://www.reuters.com/world/european-court-rules-russia-was-behind-litvinenko-killing-2021-09-21/.

14   Haroon Siddique and Andrew Roth, "Russia responsible for Alexander Litvinenko death, European court rules," *The Guardian*, September 21, 2021, https://www.theguardian.com/world/2021/sep/21/russia-responsible-for-alexander-litvinenko-death-european-court-rules.

15   Luke Harding, "Alexander Litvinenko and the most radioactive towel in history," *The Guardian*, March 6, 2016, https://www.theguardian.com/world/2016/mar/06/alexander-litvinenko-and-the-most-radioactive-towel-in-history.

16   Ibid.

17   Ibid.

the final effort to kill Litvinenko was widespread in the hotel meeting room, kitchen (the hotel even washed and reused Litvinenko's contaminated teapot), and bathroom, where one of the assassins dumped the remaining polonium.[18] Ultimately, scientists found traces of polonium across dozens of locations, including hotels, offices, a soccer stadium, and three airplanes, and tested hundreds of people for polonium poisoning.[19]

VX: KIM JONG-NAM

In 2017, North Korean operatives assassinated Kim Jong-nam, the half-brother of North Korean dictator Kim Jong-un, with the nerve agent VX at Kuala Lumpur International Airport in Malaysia.[20] Two women in their 20s, one after the other, smeared substances on Kim's face that absorbed into his skin and caused his death less than 20 minutes later. At least four North Korean agents were reportedly in the vicinity to witness the murder and step in if needed. The women, from Indonesia and Vietnam, were trained by North Korean agents to swab Kim's face and then wash their hands, according to Malaysian authorities. The event was captured on CCTV footage at the airport, and the women claimed they were carrying out a prank for a Japanese YouTube show and were unaware that they were playing a role in Kim's murder.[21] At the time of his death, Kim was carrying dozens of vials of atropine, an antidote to poisons, including VX.[22] The North Korean government denied any involvement in the incident.

Kim Jong-nam had lived in exile in Macau, a special administrative region of China, since around 2003.[23] South Korean lawmakers believed Kim Jong-un had a standing order to execute his half-brother, who publicly criticized his family's rule. Kim Jong-nam, the oldest child of Kim Jong-il, embarrassed his family in 2001 after trying to enter Japan with a forged passport to visit Tokyo Disneyland, which reportedly ended any chance for him to rule North Korea. Nonetheless, his heritage may have remained a threat in the eyes of Kim Jong-un since Kim Jong-nam remained connected to his uncle, who became the second-highest ranking North Korean after Kim Jong-il died in 2011 until his own execution in 2013. Kim Jong-nam openly denounced his half-brother and was the target of previous assassination attempts.[24]

The Malaysian government investigated the assassination and identified the perpetrators and substance used. VX is banned under the CWC, although North Korea is one of the few states not party to the treaty and reportedly has stocks of various CWs. VX is usually in a viscous liquid form, similar

18   Harding, "Alexander Litvinenko: the man who solved his own murder."

19   Steve Boggan, "Who else was poisoned by polonium?," *The Guardian*, June 4, 2007, https://www.theguardian.com/world/2007/jun/05/russia.science.

20   Heather Nauert, "Imposition of Chemical and Biological Weapons Control and Warfare Elimination Act Sanctions on North Korea," U.S. Department of State Press Statement, March 6, 2018, https://2017-2021.state.gov/imposition-of-chemical-and-biological-weapons-control-and-warfare-elimination-act-sanctions-on-north-korea/index.html.

21   James Griffiths and Salhan Ahmad, "Kim Jong Nam had antidote to VX nerve agent on him at time of murder," CNN, December 1, 2017, https://www.cnn.com/2017/11/30/asia/kim-jong-nam-antidote-intl/index.html; Hannah Ellis-Petersen and Benjamin Haas, "How North Korea got away with the assassination of Kim Jong-nam," *The Guardian*, April 1, 2019, https://www.theguardian.com/world/2019/apr/01/how-north-korea-got-away-with-the-assassination-of-kim-jong-nam; and A. Ananthalakshmi and Liz Lee, "VX used in airport murder of Kim Jong Nam kills in minutes," Reuters, February 27, 2017, https://www.reuters.com/article/us-northkorea-malaysia-kim-nerveagent/vx-used-in-airport-murder-of-kim-jong-nam-kills-in-minutes-idUSKBN1630F4.

22   Reuters Staff, "Kim Jong Nam had nerve agent antidote in bag, Malaysian court told," Reuters, December 1, 2017, https://www.reuters.com/article/uk-northkorea-malaysia-kim-court/kim-jong-nam-had-nerve-agent-antidote-in-bag-malaysian-court-told-idUKKBN1DV3TS?edition-redirect=uk.

23   James Griffiths, "Kim Jong Nam: Why would North Korea want him dead?," CNN, February 22, 2017, https://www.cnn.com/2017/02/20/asia/kim-jong-nam-north-korea/index.html.

24   Ibid; Griffiths and Ahmad, "Kim Jong Nam had antidote to VX nerve agent on him at time of murder"; Griffiths, "Kim Jong Nam: Why would North Korea want him dead?"; and "Kim Jong Nam had nerve agent antidote in bag, Malaysian court told," Reuters.

in texture to honey. It is highly toxic when it touches skin, and "a microscopic amount" could kill one person. To survive a VX attack, the victim would need an antidote administered almost immediately. The two attackers survived unharmed (although one woman vomited after) likely because VX is slow to evaporate. This means people are less likely to breathe it in, and the women washed their hands right after, an effective way to decontaminate their skin.[25]

NOVICHOK: SERGEI AND YULIA SKRIPAL

In 2019, former Russian intelligence officer Sergei Skripal and his daughter, Yulia Skripal, were poisoned with a Novichok nerve agent in an assassination attempt in Salisbury, England. They were discovered on a park bench, hallucinating and slumped over, not far from Skripal's home, where investigators found Novichok on the front door handle. Yulia was visiting to seek her father's blessing to marry her longtime boyfriend.

The United Kingdom recruited Skripal as a spy in the mid-1990s, after which he was caught by Russia and later released through a spy swap with the United States in 2010. He is of the same age and received the same training as Vladimir Putin, and Putin once admitted he "publicly daydreamed" about the death of Skripal and the other spies freed in the swap.[26] British investigations determined that the assassination operation was "almost certainly" approved at a high level in the Russian government.[27] British officials accused Russia of sending two men (and later three), all members of Russia's military intelligence unit, to kill Skripal, which Russia has repeatedly denied.[28] Russia claimed that it eliminated its entire CW stockpile in 2017. (The United States is nearly finished destroying its stockpile, a process that has taken decades.)

The Russian agents carried the Novichok in a modified perfume bottle that they sprayed on the front door handle of Skripal's home. According to the United Kingdom, the Soviet Union first developed Novichok nerve agents in the 1980s. Russia produced and stockpiled small quantities of it, even after signing the CWC, and ran a program to test delivering nerve agents, including targeting door handles.[29] Novichok can work quickly, potentially within 30 seconds with a high enough dose, and it can be absorbed through the skin, inhaled, or ingested.[30]

The Russian perpetrators were extremely careless in their attempt to kill Skripal. For example, they left the perfume bottle, filled with enough Novichok to potentially kill thousands of people, reportedly in a trash can. In nearby Amesbury, England, a local, Charlie Rowley, found the bottle, which was labeled with a known perfume brand, and gifted it to his partner, Dawn Sturgess. She sprayed the substance onto her wrists and died one week later, four months after the Skripal poisoning. Rowley was in the hospital for three weeks and survived.[31] A police officer also became seriously ill after coming into contact with the Novichok found on Skripal's front door.[32]

25   Alexander Smith, "North Korea's Kim Jong Nam Killed With VX, the 'Most Toxic Weapon Ever'," NBC News, February 24, 2017, https://www.nbcnews.com/news/north-korea/north-korea-s-kim-jong-nam-killed-vx-most-toxic-n725131.

26   Michael Schwirtz and Ellen Barry, "A Spy Story: Sergei Skripal Was a Little Fish. He Had a Big Enemy." *New York Times*, September 9, 2018, https://www.nytimes.com/2018/09/09/world/europe/sergei-skripal-russian-spy-poisoning.html.

27   Theresa May, "PM statement on the Salisbury investigation: 5 September 2018," Prime Minister's Office, September 5, 2018, https://www.gov.uk/government/speeches/pm-statement-on-the-salisbury-investigation-5-september-2018.

28   "Salisbury & Amesbury Investigation," UK Counter Terrorism Policing, September 21, 2021, https://www.counterterrorism.police.uk/salisbury/.

29   May, "PM statement on the Salisbury investigation: 5 September 2018."

30   Miriam Berger, "What is Novichok, the nerve agent linked to the Alexei Navalny poisoning?," *Washington Post*, September 24, 2020, https://www.washingtonpost.com/world/2020/08/26/what-are-chemicals-doctors-say-may-have-been-used-poison-alexei-navalny/.

31   "Novichok: Victim found poison bottle in branded box," BBC, July 24, 2018, https://www.bbc.com/news/uk-44947162.

32   Jane Corbin, "Skripal poisoning: Policeman's family 'lost everything' because of Novichok," BBC Panorama, November 22, 2018, https://www.bbc.com/news/uk-46290989.

NOVICHOK: ALEXEI NAVALNY

In 2020, Russian agents used a Novichok nerve agent to poison Russian opposition leader Alexei Navalny at a hotel in a Siberian city Tomsk in Russia, where he was meeting with opposition candidates. Navalny, often deemed Putin's "most persistent critic," became sick and collapsed 30 minutes into a flight, which made an emergency landing in another Siberian city, Omsk. He was hospitalized and flown to Berlin two days later. The doctors in Omsk claimed Navalny was suffering from low blood sugar, while the German doctors knew it was more serious and reportedly requested assistance from military scientists with expertise in nerve agents. The doctors even placed Navalny in a medically induced coma for several weeks.[33]

A German military laboratory, two independent European laboratories, and the OPCW all found that the poison used against Navalny was in the Novichok family. Investigative journalism group Bellingcat identified three Russian operatives, two of which were medical doctors, who tailed Navalny and at least five others who were involved.[34] Bellingcat discovered that Navalny had been surveilled for years by Russian operatives. Putin claimed the tails were needed because Navalny was working for U.S. intelligence, which Navalny denied.[35]

After he recovered, Navalny posed as an aide to a top Russian general in a phone call to one the operatives involved in his assassination attempt. As published by Bellingcat, Navalny learned that Russian agents poisoned him by applying Novichok to the inside of his boxer shorts at his hotel. In the phone call, the Russian operative said he retrieved and cleaned the clothes Navalny wore when he was poisoned to destroy evidence of the crime. Navalny survived because of the time it took for the Novichok to transfer from his clothes to his skin and take effect and due to the emergency landing, after which they removed his clothes, gave him an antidote, and put him on a ventilator.[36] While this is the only attack of the four highlighted that occurred on the perpetrator's domestic territory, moving Navalny to a German hospital facilitated the international community's ability to recognize and investigate the incident.

MOTIVATIONS AND CHALLENGES WITH USING WMDS IN ASSASSINATIONS

While the true reason for using WMDs in assassinations or attempted assassinations in each case will likely remain unknown to the public, the four cases reviewed offer trends that can point to theories. WMD use is notable in these examples for the associated challenges with each operation and the opportunity to provide plausible deniability. Despite the uncertainty of an attack's success, the potential for contamination, and the ability to identify substances and perpetrators, Russia and North Korea have found WMD poisons to be an appealing weapon of choice for assassinations, particularly on foreign

---

33   Michael Schwirtz and Melissa Eddy, "Aleksei Navalny Was Poisoned With Novichok, Germany Says," *New York Times*, September 2, 2020, Updated April 30, 2021, https://www.nytimes.com/2020/09/02/world/europe/navalny-poison-novichok.html.

34   Bellingcat Investigation Team, "FSB Team of Chemical Weapon Experts Implicated in Alexey Navalny Novichok Poisoning," Bellingcat, December 14, 2020, https://www.bellingcat.com/news/uk-and-europe/2020/12/14/fsb-team-of-chemical-weapon-experts-implicated-in-alexey-navalny-novichok-poisoning/.

35   Luke Harding, "Navalny says Russian officer admits putting poison in underwear," *The Guardian*, December 21, 2020, https://www.theguardian.com/world/2020/dec/21/navalny-russian-agent-novichok-death-plot.

36   Harding, "Navalny says Russian officer admits putting poison in underwear"; and Bellingcat Investigation Team, "'If it Hadn't Been for the Prompt Work of the Medics': FSB Officer Inadvertently Confesses Murder Plot to Navalny," Bellingcat, December 21, 2020, https://www.bellingcat.com/news/uk-and-europe/2020/12/21/if-it-hadnt-been-for-the-prompt-work-of-the-medics-fsb-officer-inadvertently-confesses-murder-plot-to-navalny/.

soil. Of the four cases reviewed, only two were "successful," resulting in the target's death. Navalny and the Skripals survived due to relatively quick medical intervention and perhaps a failure to absorb the intended dose. Litvinenko's death was drawn out over weeks, giving time for him to work with investigators and share his story with the world. Finally, Kim Jong-nam died within minutes, which may make his assassination the "most successful" in the eyes of states ordering assassinations.

Not only are WMD poisons unreliable in terms of their level of "success," as demonstrated by the outcome of the four cases, but they also can be challenging to handle and administer. In any instance of a WMD assassination, the nature of the weapon creates the potential for unintended and dangerous contamination. The UK government's investigation into the attack on Litvinenko documented this well, revealing trails of polonium across London and on airplanes over three assassination attempts. Given the threat to public safety, more than 700 people were tested for polonium poisoning.[37] In the case of the Skripals, the attackers left behind a nicely packaged perfume bottle of Novichok agent, which did ultimately kill a British national and make two others seriously ill—and which could have killed thousands.

Another aspect that has failed to deter WMD assassins is the ability of forensic investigations to identify perpetrators and analyze poisons for clues. Attackers are often caught on camera. CCTV footage played a significant role in tracking the assassins in Litvenenko's case.[38] Detectives went through more than 11,000 hours of CCTV and identified two Russian men directly responsible for the Skripal assassination attempt who were placed "in the immediate vicinity of the Skripals' house . . . moments before the attack."[39] In the case of Kim Jong-nam, airports are one of the most surveilled public spaces, suggesting that North Korea knew the world would find out about the attack. The director of a documentary on the subject, whose team spent three months going through thousands of hours of footage, believed "Kim Jong-un and those working for him wanted a very public murder to show the whole world what happens when people displease the nation's supreme leader or get in his way. Even if they're family."[40] Bellingcat's groundbreaking open-source investigations, including surrounding Russia's "poison squad," have been successful by tracking not just camera footage but also information such as flight records and phone metadata.

Furthermore, laboratories can analyze the poisons, whether through blood and urine samples of those harmed or samples of the substance itself, to identify them and potentially determine their origin. Polonium-210 and the VX and Novichok nerve agents all indicate state-sponsored attacks, and government and OPCW-designated laboratories were able to pinpoint the poison in each case. Polonium-210 is only available publicly in very small quantities and "difficult to produce and dangerous to handle." Furthermore, based on analyses, the substance used for Litvinenko's assassination most likely came from a Russian nuclear reactor (and would have cost tens of millions of dollars to purchase based on the quantity and purity).[41] VX is challenging to produce and requires an advanced laboratory.[42]

In the case of the Novichok attacks, the United States "stopped producing nerve agents in 1970, after the development of 'third generation' nerve agents like sarin and VX. Soviet scientists kept at it for

---

37 Boggan, "Who else was poisoned by polonium?"
38 Harding, "Alexander Litvinenko: the man who solved his own murder."
39 May, "PM statement on the Salisbury investigation: 5 September 2018."
40 Dowd, "Assassins: How CCTV gave Kim Jong-nam murder documentary added intrigue."
41 Owen, *The Litvinenko Inquiry*, 216–17, 225–26.
42 Smith, "North Korea's Kim Jong Nam Killed With VX, the 'Most Toxic Weapon Ever.'"

two more decades, developing a 'fourth generation,' the Novichok group of weapons."[43] The Skripal poisoning "turned Novichok into something of a Russian calling card."[44] Bellingcat reported, based on the OPCW's determination, that the Novichok that poisoned Navalny structurally resembled known Novichok variants but was not the same as the one used in the Skripal poisoning, which "implied that the agent used on Navalny was of a more recent, previously unknown type."[45]

Putin's administration has become known for attacks using poisons in an attempt to silence or kill opponents, not to mention using more "traditional" means such as guns or a push from a tall building, that fit into Russia's long history of assassinating political dissidents.[46] While conventional means for murder tend to be more straightforward, WMD poisons lend an air of ambiguity and a layer of separation since they can be consumed in beverages or rubbed on skin and have a delayed impact on the victim. Navalny and the Skripals were poisoned through contact with their own belongings, so the operatives had time to be a safe distance away.[47] While they waited in the wings at the airport, North Korean operatives arranged for seemingly innocent assassins without ties to North Korea to carry out the murder, adding to the ambiguity surrounding the case. The operatives then quickly boarded planes out of Malaysia, and their flight routes avoided countries that might ground their planes and arrest them. One report suggested that North Korea recruited two innocent women "to avoid a repeat of the Rangoon bombing incident in Burma in 1983, when two North Korean officers who attempted to publicly assassinate the South Korean president . . . were captured and put on trial."[48]

However, perhaps Russian and North Korean leadership appreciates some of the publicity surrounding the WMD poisonings, providing a level of prestige as the world learns they possess these weapons. The incidents also send a clear message aimed at silencing critics and other perceived threats, which was successful with prominent North Korean defectors, at least in part. The former North Korean ambassador to the United Kingdom and several others who had been speaking publicly "cancelled all public appearances" and "kept a low profile" following Kim Jong-nam's assassination. A North Korea expert speculated that the public nature of his murder was "to leave a calling card, to show the world that Kim Jong-un is not afraid to use a weapon of mass destruction at a crowded international airport."[49] Russian agents used Litvinenko's murder as an apparent threat to another Putin critic and associate of Litvinenko, Boris Berezovsky, by handing him a t-shirt that said "POLONIUM-210 CSKA LONDON, HAMBURG To Be Continued" on the front and "CSKA Moscow Nuclear Death Is Knocking Your Door" on the back.[50] Russia's use of Novichok agents in assassination attempts, according to one journalist, is "about telling Western governments not to tangle with Russia" and to send "a message with a very clear return address."[51]

INTERNATIONAL RESPONSE FOR ACCOUNTABILITY

In the aftermath of the four WMD poisoning cases outlined, states responded unilaterally and through coordinated efforts with international and multilateral organizations to condemn the attacks, impose economic and diplomatic punishments, engage in forensic investigations and

43   Schwirtz and Eddy, "Aleksei Navalny Was Poisoned With Novichok, Germany Says."
44   Ibid.
45   Bellingcat Investigation Team, "FSB Team of Chemical Weapon Experts Implicated in Alexey Navalny Novichok Poisoning."
46   Owen, *The Litvinenko Inquiry*, 231, 233.
47   Tom Stevenson, "Why use a Novichok?," *London Review of Books* 43, no. 9 (May 2021), https://www.lrb.co.uk/the-paper/v43/n09/tom-stevenson/why-use-a-novichok.
48   Ellis-Petersen and Haas, "How North Korea got away with the assassination of Kim Jong-nam."
49   Ibid.
50   Owen, The Litvinenko Inquiry, 199–202.
51   Madeline Roache, "Russia Was Linked to 14 Deaths in the U.K., but Britain Looked Away. A New Book Explores Why," *TIME*, November 12, 2019, https://time.com/5725041/uk-russian-assassins-heidi-blake/.

court cases, and update international mechanisms to hold perpetrators accountable (see Table 2). Three of four attacks occurred on foreign soil, straining or further damaging relationships between named perpetrators and accusers. In terms of legal and economic repercussions, most focused on the individuals involved in the operation rather than a state entity. Diplomatic responses such as expelling diplomats contributed to further isolating North Korea and Russia. Naming and shaming played a major role to make clear that violating international norms and taboos is problematic, although this only goes so far in hurting countries that are already facing diplomatic, economic, and legal consequences. The Skripal poisoning was a clear turning point, resulting in additions to the CWC Annex on Chemicals and the Australia Group control list and giving the OPCW greater authority to attribute CW attacks through the use of forensic tools and techniques.

Table 2: Snapshot of Responses by States and International Organizations

	Diplomatic/Political	Economic	Legal/Authority	Change in Designation of Poison Used
RUSSIA **Alexander Litvinenko**	– Name and shame: many states – Expel diplomats/agents: United Kingdom – Suspend expedited visa processing: United Kingdom	– Sanctions: United Kingdom, United States	– Investigations: United Kingdom, Germany, European Court of Human Rights – Warrants for arrest: United Kingdom – Court cases: European Union	– Considered but no clear change by International Atomic Energy Agency
NORTH KOREA **Kim Jong-nam**	– Name and shame: many states – Expel diplomats/agents: Malaysia	– Sanctions: United States	– Investigations: Malaysia, Organization for the Prohibition of Chemical Weapons – Warrants for arrest: Malaysia – Court cases: Malaysia	– None; (already banned under Chemical Weapons Convention
RUSSIA **Sergei and Yulia Skripal**	– Name and shame: many states – Expel diplomats/agents: many states, in a coordinated response	– Sanctions: European Union, United States	– New authority for Organization for the Prohibition of Chemical Weapons to attribute perpetrator in chemical attacks	– Expanded list of banned chemicals under Chemical Weapons Convention – Expanded Australia Group control list of Chemical Weapons precursors
RUSSIA **Alexei Navalny**	– Name and shame: many states	– Sanctions: United Kingdom, United States	– Investigations: Germany, Organization for the Prohibition of Chemical Weapons, France, Sweden	– None

Note: This table is a snapshot of how states and international organizations responded to the attacks, with a particular focus on the United Kingdom, United States, Germany, Malaysia, European Union, and the Organization for the Prohibition of Chemical Weapons (OPCW), as relevant to each incident. It is not intended to be comprehensive.

Source: Author's research and analysis.

In response to the assassination of Alexander Litvinenko, the United Kingdom, United States, and others named and shamed Russia. The United Kingdom expelled four Russian diplomats and suspended a new effort to provide visa application processing for Russian citizens.[52] The United Kingdom and United States also imposed sanctions against Russian individuals involved in the operation.[53] The United Kingdom conducted police investigations, including forensic analysis, and issued warrants for the Russian agents' arrest. Germany also conducted a police investigation after discovering polonium that may have been tied to the Russian perpetrators.[54] The ECHR brought forth its own case on the matter and found Russia responsible. Additionally, the International Atomic Energy Agency (IAEA) considered changing its designation of polonium-210 from "unlikely to be dangerous," category four on a five-tier scale, to a higher tier for more dangerous radioactive sources.[55] The United Kingdom also added the Litvinenko poisoning to the IAEA's Incident and Trafficking Database.[56]

After Kim Jong-nam's assassination with VX in Malaysia, the United States and others condemned North Korea. Malaysia avoided naming North Korea directly, expressing its strong condemnation for "the use of such a chemical weapon by anyone, anywhere and under any circumstances."[57] Unlike the Russian cases, North Korea and Malaysia had decades of strong ties before the assassination of Kim Jong-nam. However, the incident created a diplomatic rift in which Malaysia expelled North Korea's ambassador, demanded that three North Koreans hiding in the embassy be questioned by police, and refused to release Kim Jong-nam's body. In turn, North Korea banned Malaysians in North Korea from leaving the country.[58] Malaysia eventually let the three North Koreans return home, released the body, and said it was not breaking diplomatic ties with North Korea, but it planned to continue the murder investigation and enlisted technical assistance from the OPCW.[59] Malaysia proceeded with the murder trial against the two women, largely viewed as pawns, and both were ultimately released after spending time in jail.[60] In addition, the United States sanctioned North Korean individuals involved

52   Mark Tran, "Britain expels four Russian diplomats," *The Guardian*, July 16, 2007, https://www.theguardian.com/uk/2007/jul/16/russia.politics.

53   Catherina Valenzuela-Bock, "U.K. Announces Sanctions on Those Deemed Responsible for Litvinenko Murder (January 21, 2016)," American Society of International Law, February 1, 2016, https://www.asil.org/blogs/uk-announces-sanctions-those-deemed-responsible-litvinenko-murder-january-21-2016; Office of Foreign Assets Control, "Sanctions Actions Pursuant to the Sergei Magnitsky Rule of Law Accountability Act of 2012," Federal Register, January 13, 2017, https://www.federalregister.gov/documents/2017/01/13/2017-00603/sanctions-actions-pursuant-to-the-sergei-magnitsky-rule-of-law-accountability-act-of-2012; and Shaun Walker, "Litvinenko suspects added to US sanctions list against Russia," *The Guardian*, January 10, 2017, https://www.theguardian.com/us-news/2017/jan/10/litvinenko-suspects-added-to-us-russia-sanctions-list.

54   Louis Charbonneau, "Germany tracks polonium trail left by ex-spy contact," Reuters, January 20, 2007, https://www.reuters.com/article/us-britain-poisoning/germany-tracks-polonium-trail-left-by-ex-spy-contact-idUSL1168128220061211.

55   Mark Heinrich, "IAFA reviews polonium risk," Reuters, January 20, 2007, https://www.reuters.com/article/uk-iaea-polonium/iaea-reviews-polonium-risk-idUKL1332839820061213.

56   Christopher Ashley Ford, "The Challenge and the Potential of U.S.-Russian Nonproliferation Cooperation," U.S. Department of State, October 22, 2018, https://2017-2021.state.gov/remarks-and-releases-bureau-of-international-security-and-nonproliferation/the-challenge-and-the-potential-of-u-s-russian-nonproliferation-cooperation/index.html.

57   Richard C. Paddock and Choe Sang-Hun, "Use of Nerve Agent in Kim Jong-nam Killing Is Condemned by Malaysia," *New York Times*, March 2, 2017, https://www.nytimes.com/2017/03/02/world/asia/kim-jong-nam-malaysia.html.

58   Ellis-Petersen and Haas, "How North Korea got away with the assassination of Kim Jong-nam."

59   Rozanna Latiff and James Pearson, "North Korean murder suspects go home with victim's body as Malaysia forced to swap," Reuters, March 30, 2017, https://www.reuters.com/article/us-northkorea-malaysia-kim/north-korean-murder-suspects-go-home-with-victims-body-as-malaysia-forced-to-swap-idUSKBN1711NI; "Malaysia: Statement at the 84th session of the Executive Council," OPCW, March 7, 2017, https://www.opcw.org/ec-84; and "Decision: Chemical Weapons Incident in Kuala Lumpur, Malaysia," OPCW Executive Council, March 9, 2017, https://www.opcw.org/sites/default/files/documents/EC/84/en/ec84dec08_e_.pdf.

60   Vincent Dowd, "Assassins: How CCTV gave Kim Jong-nam murder documentary added intrigue," BBC News, February 8, 2021, https://www.bbc.com/news/entertainment-arts-55940440.

in the assassination and cited this event as one reason why it re-designated North Korea as a state sponsor of terrorism.[61]

Following the Skripals' Novichok poisoning, a slew of countries named and shamed Russia, with the G7 and other groups of countries releasing joint statements.[62] In a coordinated effort, 26 countries expelled 143 Russian intelligence officers, and NATO expelled seven Russian officials.[63] The European Union, including the United Kingdom, imposed sanctions on Russians involved in the attack.[64] The United States imposed multiple rounds of sanctions.[65] The United Kingdom conducted a forensic investigation with technical assistance from the OPCW.[66] It later charged two Russian men in absentia with conspiracy to murder and attempted murder and named a third suspect.[67] In addition to the OPCW providing technical assistance to the United Kingdom, it also achieved two highly notable changes after the Skripals' poisoning. First, in 2018, CWC member states voted to provide the OPCW with new authority to attribute chemical attacks under investigation to a perpetrator, organizer, or sponsor.[68] Second, in 2019, CWC member states updated the list of Schedule 1 chemicals, banned as chemical warfare agents, to include two classes of Novichok nerve agents and some other chemical compounds.[69] While the use of any chemical as a weapon was already prohibited under the CWC, this was the first time in history that members updated the list. The Australia Group, a multilateral export control regime, also updated its control list to include Novichok precursors.[70]

Similarly, the United States, EU and NATO leaders, Germany, and other European countries condemned Russia for the Novichok attack on Navalny.[71] The United States imposed multiple rounds and types of

61   Nauert, "Imposition of Chemical and Biological Weapons Control and Warfare Elimination Act Sanctions on North Korea"; and Michael D. Shear and David E. Sanger, "Trump Returns North Korea to List of State Sponsors of Terrorism," *New York Times*, November 20, 2017, https://www.nytimes.com/2017/11/20/us/politics/north-korea-trump-terror.html.

62   "Novichok nerve agent use in Salisbury: UK government response, March to April 2018," UK Government, last updated April 18, 2018, https://www.gov.uk/government/news/novichok-nerve-agent-use-in-salisbury-uk-government-response; "Salisbury attack: Joint statement from the leaders of France, Germany, the United States and the United Kingdom," UK Government, March 15, 2018, https://www.gov.uk/government/news/salisbury-attack-joint-statement-from-the-leaders-of-france-germany-the-united-states-and-the-united-kingdom; and "Joint Statement on the Salisbury Attack," U.S. Department of State, September 6, 2018, https://2017-2021.state.gov/joint-statement-on-the-salisbury-attack/index.html.

63   "Novichok nerve agent use in Salisbury," UK Government.

64   "EU imposes sanctions against Salisbury suspects," UK Government, January 21, 2019, https://www.gov.uk/government/news/eu-imposes-sanctions-against-salisbury-suspects.

65   Cory Welt, Kristin Archick, Rebecca M. Nelson, and Dianne E. Rennack, *U.S. Sanctions on Russia*, CRS Report No. R45415 (Washington, DC: Congressional Research Service, updated January 2022), 20–21, https://crsreports.congress.gov/product/pdf/R/R45415.

66   "OPCW Issues Report on Technical Assistance Requested by the United Kingdom," OPCW, April 12, 2018, https://www.opcw.org/media-centre/news/2018/04/opcw-issues-report-technical-assistance-requested-united-kingdom; and Ambassador Karen Pierce, "Evidence of Russia's Involvement in Salisbury Attack," UK Government, September 6, 2018, https://www.gov.uk/government/speeches/you-dont-recruit-an-arsonist-to-put-out-a-fire-you-especially-dont-do-that-when-the-fire-is-one-they-caused.

67   Scott Neuman, "Russia Fatally Poisoned A Prominent Defector In London, A Court Conclude," NPR, September 22, 2021, https://www.npr.org/2021/09/21/1039224996/russia-alexander-litvinenko-european-court-human-rights-putin.

68   "CWC Conference of the States Parties Adopts Decision Addressing the Threat from Chemical Weapons Use," OPCW, June 27, 2018, https://www.opcw.org/media-centre/news/2018/06/cwc-conference-states-parties-adopts-decision-addressing-threat-chemical.

69   Mary Beth D. Nikitin, "Resurgence of Chemical Weapons Use: Issues for Congress," Congressional Research Service, updated March 5, 2021, https://crsreports.congress.gov/product/pdf/IN/IN10936/12.

70   "Statement by the Australia Group Chair: Addition of Novichok precursor chemicals to the Australia Group Control List," Australia Group, February 28, 2020, https://www.dfat.gov.au/publications/minisite/theaustraliagroupnet/site/en/statement-by-the-australia-group-chair-addition-of-novichok-precursor-chemicals-to-the-australia-group-control-list.html.

71   Pierre Morcos, "The Implications of the Poisoning of Alexey Navalny," CSIS, *Critical Questions*, September 10, 2020, https://www.csis.org/analysis/implications-poisoning-alexey-navalny.

sanctions on Russia and increased limitations on defense-related exports.[72] The United Kingdom also imposed multiple rounds of sanctions.[73] Germany investigated the incident and requested technical assistance from the OPCW, France, and Sweden.[74] Using a formal mechanism, Article VII under the CWC, a group of 45 states parties sent questions to Russia seeking clarification on the Navalny poisoning and how Russia was investigating the use of a Novichok nerve agent in its own country. Russia did not respond to the questions but rather countered with its own questions, after which 55 states parties issued a statement urging Russia to respond to the original questions.[75] Russia has yet to respond.

IMPACT FOR GLOBAL NORMS AND TABOOS

Russia and North Korea have used WMD poisons to assassinate or attempt to assassinate political enemies despite the challenges associated with highly toxic substances that violate international treaties and norms and degrade long-standing taboos against WMD use. Perpetrators appear to use these substances because they favor their ambiguous nature, opening a path for misinformation and disinformation in the international sphere. While these autocratic regimes seem to prefer clandestine methods and will deny attribution, even if they are caught through the variety of investigative methods available today, their message reverberates around the world: using whatever means are necessary, they are willing to kill anyone who threatens the status quo. Despite the condemnation that follows, revealing possession of a dangerous WMD poison may also provide a status boost, making other states take the perpetrator more seriously as a threat on the world stage, even as they publicly deny involvement.

Any use of a WMD erodes norms for behavior and taboos against their use and is highly concerning. The four highlighted cases demonstrate the erosion but not the elimination of norms against CW use and taboos against using radiological weapons. Since the mid-2000s, North Korea and Russia have used WMD assassinations for a narrow scope of silencing dissidents who are one of "their own," and this has not correlated with use of the same weapons for purposes of inflicting mass casualties. Only these two countries are known to use WMDs for assassinations, and their frequency is limited. Radiological weapons have not been used for assassination in any public way since Litvinenko's death, although, comparatively, CW assassinations have been on the rise. Notably, these norms and taboos have been resilient in preventing widespread use, in particular as democracies continue to demand accountability following WMD attacks.

However, it is important to note that both Russia and North Korea have contributed to the erosion of WMD norms against use outside of the realm of assassinations. Russia used a CW to incapacitate Chechen rebels who took several hundred Moscow theater-goers hostage in 2002; the government's rescue attempt ended up inadvertently killing at least 170 innocent people. Russia has enabled Syria's CW use in its civil war, and the United Nations found that North Korea provided assistance to Syria's CW program.[76] Additionally, the United States warned that Russia might use CWs in its 2022

---

72   Welt et al., *U.S. Sanctions on Russia*, 21–23.

73   PA Media, "UK imposes sanctions on seven Russians over Navalny poisoning," *The Guardian*, August 20, 2021, https://www.theguardian.com/world/2021/aug/20/uk-imposes-sanctions-seven-russians-alexei-navalny-poisoning.

74   Steffen Seibert, "Statement by the Federal Government on the Navalny case," Government of Germany, September 14, 2020, https://www.bundesregierung.de/breg-en/news/statement-by-the-federal-government-on-the-navalny-case-1786624.

75   Hanna Notte, "Chemical Weapons Impasse Reflects Russia's Broader Conflict With the West," Carnegie Moscow Center, December 16, 2021, https://carnegiemoscow.org/commentary/86015.

76   Michael Schwirtz, "U.N. Links North Korea to Syria's Chemical Weapons Program," *New York Times*, February 27, 2018,

invasion of Ukraine.[77] Despite these incidents, Russia and North Korea have not directly used CWs, BWs, or radiological or nuclear weapons to incite mass casualties.

WMD assassinations appear most likely to be another tool in the toolbox for autocracies to achieve their political goals of silencing dissidents and discouraging others from speaking out against the regime. Dictators favor this type of repression as a tool for maintaining leadership control and implementing their broader security strategy, including committing all types of assassinations, silencing media that is not aligned with the state, putting opposition candidates (e.g., Navalny) in jail, skewing elections or exerting total control over the political system, and other systemic activities to control the narrative and maintain power. Suppressing these apparent threats to power and sending a clear message domestically may also fit in with broader geopolitical goals for an autocracy by limiting domestic interference and freeing up capacity and resources to, for example, invade a neighboring country.[78]

Democracies' efforts to name and shame, reduce diplomatic ties, sanction, and prosecute offenders of laws and norms against the use of WMDs are important to counter autocracies' narratives and demonstrate that their efforts to push back against systems for international security will not go unchecked. Domestically, internal accountability is unrealistic given limitations on the information environment. Autocracies continue to push boundaries to undermine international institutions to serve their interests. According to Freedom House, "Only global solidarity among democracy's defenders can successfully counter the combined aggression of its adversaries."[79] Even if autocracies continue to commit WMD assassinations, unilateral and coordinated responses from democracies are meaningful to make clear to each other and the world that violations of treaties, norms, and taboos continue to be unacceptable in the rules-based global order.

CONCLUSION

Despite international efforts to hold states accountable for WMD assassinations and assassination attempts, it appears that these kinds of WMD attacks may continue as long as the benefits to the state outweigh the costs of the international response. Perpetrators are likely learning from their missteps and testing perceived limits to see how strongly the international community will react. If this is the case, norms and taboos will continue to erode with these incidents, most strongly if new countries start ordering WMD assassinations on the international stage. At this point, the use of WMDs for assassinations could still be contained if democracies act to counter autocratic leaders in ways that are meaningful both to the offender and to observing states. North Korea is an unusual case because it is already heavily isolated, shamed, and sanctioned for its behavior violating international norms, and Russia is increasingly going down that path, particularly with its most recent invasion of Ukraine. The difficulty in developing WMD poisons adds another barrier to entry that states and international organizations can manage through nonproliferation tools such as export controls, safeguards, and verification regimes. The limited use of WMDs for assassinations by

https://www.nytimes.com/2018/02/27/world/asia/north-korea-syria-chemical-weapons-sanctions.html.

77   Zeke Miller, "White House warns Russia may use chemical weapons in Ukraine," Associated Press, March 9, 2022, https://apnews.com/article/russia-ukraine-europe-jen-psaki-chemical-weapons-weapons-of-mass-destruction-f01cedf434ec-697034b2912696a3448b.

78   Daniel Treisman, "Putin Unbound: How Repression at Home Presaged Belligerence Abroad," *Foreign Affairs*, May/June 2022, https://www.foreignaffairs.com/articles/ukraine/2022-04-06/putin-russia-ukraine-war-unbound.

79   Sarah Repucci and Amy Slipowitz, *Freedom in the World 2022: The Global Expansion of Authoritarian Rule* (Washington, DC: Freedom House, February 2022), https://freedomhouse.org/report/freedom-world/2022/global-expansion-authoritari-an-rule.

select autocracies reinforces the bigger picture that almost all states are *not* using WMDs for these purposes or at all. While the international system is not set up to counter a state willing to take certain actions regardless of the consequences, it can at least prevent and deter others who wish to maintain a positive international image and follow the rules that underpin global peace and security.

Trust the Process

Assessing the Value of Future U.S.-Russian Cyber-Nuclear Diplomacy for Strategic Stability

By Dan Zhukov[1]

INTRODUCTION

Since the advent of atomic weapons, the world has lived under a nuclear shadow, with nuclear weapons states relying on mutual deterrence to avert catastrophe. During the Cold War, diplomacy between these states provided an avenue for deterrence messaging and reinforced strategic stability. However, as emerging digital tools have created ways to exploit nuclear weapons' cyber vulnerabilities in recent decades, the dynamic of deterrence has grown more complex and unstable. Can diplomacy continue serving its stabilizing role and reduce the cyber-driven risks of nuclear escalation?

In a growing crisis between nuclear weapon states, cyber operations involving nuclear arsenals or infrastructure may be one of the most unpredictable and dangerous pathways toward nuclear escalation. As Herbert Lin discusses at length in his book *Cyber Threats and Nuclear Weapons*, nuclear command, control, and communications (NC3) infrastructure, early warning systems, weapons platforms, and nuclear warheads themselves all use software that may be potentially vulnerable to cyber exploits.[2] Furthermore, as long as an adversary fears its nuclear deterrent is under significant threat, efforts to exploit these vulnerabilities may lead to escalation even if they do not actually cripple the adversary's nuclear arsenal.[3]

---

1   Dan Zhukov is a research assistant at the Center for International Security and Cooperation (CISAC) at Stanford University. The views expressed in this paper are those of the author and do not reflect the views of CISAC, Stanford, or any other organizations or institutions. The author also expresses sincere thanks to the CSIS Project on Nuclear Issues team for supporting this project, to the interview subjects for their involvement and feedback, and to Rose Gottemoeller for her advice and comments on this paper.
2   Herb Lin, *Cyber Threats and Nuclear Weapons* (Stanford, CA: Stanford University Press, January 2021).
3   James Acton, "Cyber Warfare & Inadvertent Escalation," *Daedalus* 149, no. 2 (Spring 2020), doi:10.2307/48591317.

This fear could also lead nuclear weapon states, including the United States, to take steps toward dissuading their adversaries, such as Russia, from conducting cyberattacks on their nuclear enterprises, especially while modernizing their NC3 and other weapons systems. Since these operations would take place within cyberspace, they naturally fall within the U.S. policy called "defend forward"—the strategy of deterring cyber threats by hunting for their sources in adversarial cyberspace.[4] This strategy incorporates allied diplomacy as a means to bolster attribution and impose greater costs on a potential cyber adversary. And yet, less has been written within the U.S. cyber defense strategy on the role of dialogue with the adversary, especially in the nuclear context. As such, this paper examines the role bilateral diplomacy with Russia can play in the U.S. strategy to mitigate cyber risks to nuclear stability and deter malicious cyber-nuclear threats.

However, considering Russia's unprovoked aggression in Ukraine, it is difficult to foresee a successful negotiation of bilateral cyber-nuclear risk reduction and confidence-building measures between the United States and Russia in the short or medium term. Even after the war in Ukraine ends, the question of engaging in diplomacy with the current Russian government is rife with moral issues stemming from current geopolitical and humanitarian realities. To be clear, this paper will not focus on the morality of negotiating with Russia currently. Instead, it will assume that, as the countries with the world's biggest nuclear arsenals, the United States and Russia will inevitably have to discuss existential questions of strategic stability and the cyber-driven risks of nuclear escalation, no matter how much time it takes to resume the dialogue. In preparation for future negotiations, it is important to explore risk reduction techniques and measures now.

The history of nuclear arms control points to the stabilizing effect of diplomatic outcomes such as confidence-building measures and codified restraints. Per Anne Sartori's main argument in *Deterrence by Diplomacy*, diplomacy is also an inherently valuable tool of deterrence because it allows states to communicate threats that may alter their adversaries' perception.[5] Both of these functions—successful negotiations and deterrence communication—require diplomatic interlocutors to share a common base of understanding and mutual knowledge about the issues they discuss. Therefore, this paper hypothesizes that the process of bilateral cyber-nuclear diplomacy enables a shared understanding of associated risks and thus bolsters strategic stability in and of itself. Specifically, this shared understanding may lay the foundation for credible deterrence signals and formally negotiated restraints. As such, the United States should pursue bilateral dialogue on cyber-nuclear risks with Russia even if it considers concrete risk reduction measures too politically difficult to negotiate from the outset.

The rest of this paper will proceed as follows. The literature review will elaborate on the escalatory risks of cyber operations to nuclear weapons enterprises and strategic stability writ large, outline U.S. cyber deterrence policy, and then analyze different viewpoints on the role of bilateral diplomacy in this policy. The argument section will lay out the paper's thesis, seeking to synthesize the literature's complementing approaches to the stabilizing benefits of the process and results of diplomacy. The evidence section will draw on the history of the U.S.-USSR dialogue on nuclear issues during the Cold War, which enabled mutual deterrence messaging and improved strategic stability through shared learning and formal restraint measures. This section will also recount past bilateral interactions between the United States and Russia in the cyber-nuclear context, focusing strictly on the context of the dyad while recognizing the important confidence-building aspect of multilateral negotiations.

---

4   "Summary: Department of Defense Cyber Strategy," U.S. Department of Defense, 2018, https://media.defense.gov/2018/Sep/18/2002041658/-1/-1/1/CYBER_STRATEGY_SUMMARY_FINAL.PDF.

5   Anne Sartori, *Deterrence by Diplomacy* (Princeton, NJ: Princeton University Press, 2005).

The interviews section will outline the results of a series of interviews with officials and experts familiar with U.S. and Russian nuclear and cyber policies, demonstrating partial agreement with the hypothesis and outlining challenges for credible deterrence signals and restraint measures. Lastly, the conclusion will argue that U.S.-Russian cyber-nuclear dialogue will be worth pursuing after the war in Ukraine because of the dialogue's likely stabilizing effect, even if the obstacles to deterrence communication and mutual restraints cannot be overcome.

LITERATURE REVIEW

The following section will elaborate on the main factors through which cyber operations and accidents may deteriorate nuclear stability, noting the deliberate and inadvertent escalation risks associated with these operations. As a subtopic, the review will detail risks associated with malicious cyber operations targeting nuclear arsenals and describe the current U.S. policy toward deterring cyberattacks. It will then outline a range of views among Western and Russian cyber scholars and practitioners on what bilateral diplomacy measures can potentially contribute to cyber-nuclear risk reduction and nuclear stability.

Several factors make cyber operations against an adversary's nuclear weapons infrastructure particularly destabilizing. First, it is difficult to attribute a detected cyberattack to a given adversary quickly and accurately.[6] Second, it may be impossible to quickly distinguish a cyber-espionage operation from a cyberattack.[7] The 2020 SolarWinds breach is a key example of a digital breach in a part of the U.S. nuclear weapons enterprise that may have primarily constituted an espionage operation rather than a cyberattack with disruptive effects. Last but not least, cyberattacks themselves can have unpredictable effects that even the attacker cannot foresee.[8] All of these destabilizing factors are compounded by the potential use of advanced persistent threats, which the adversaries can potentially implant in the target systems or organizations well in advance of executing future operations.[9] These factors mean that a cyber operation conducted in an adversary's system governing some part of the nuclear weapons enterprise can have far-reaching and dangerous consequences. These consequences may not necessarily lead to nuclear escalation, as recent wargaming research indicates.[10] Nevertheless, such cyber operations can induce miscalculations or make the adversary lose confidence in their nuclear deterrent, creating "use-or-lose" incentives and thereby running the risk of triggering "wormhole escalation."[11]

Conversely, accidents caused by digital vulnerabilities and malfunctions may have similarly destabilizing effects without any malicious intent behind them. Technical malfunctions in the NC3, early warning, and nuclear weapons systems are not new, but the growing digitization of these systems introduces new potential pathways to failure. A technical malfunction in the U.S. and Russian nuclear weapons systems that could trigger an accidental launch was a major concern in the late

---

6   Mark Montgomery and Erica Lonergan, "Cyber Threats and Vulnerabilities to Conventional and Strategic Deterrence," *Joint orce Quarterly* 102 (2021): 82, https://www.ndu.edu/Portals/68/Documents/jfq/jfq-102/jfq-102_79-89_Features-Cyber_Threats.pdf.

7   Robert Jervis, "Some Thoughts on Deterrence in the Cyber Era," *Journal of Information Warfare* 15, no. 2 (2016): 71, https://www.jstor.org/stable/26487532?seq=7.

8   Acton, "Cyber Warfare & Nuclear Escalation."

9   "Advanced Persistent Threat (APT)," Crowdstrike, June 15, 2022, https://www.crowdstrike.com/cybersecurity-101/advanced-persistent-threat-apt/.

10   Jacquelyn Schneider, Benjamin Schechter, and Rachel Schaffer, "Cyber Operations and Nuclear Use: A Wargaming Exploration," SSRN, November 4, 2021, https://ssrn.com/abstract=3956337.

11   Rebecca Hersman, "Wormhole Escalation in the New Nuclear Age," Texas National Security Review, Summer 2020, doi:10.26153/tsw/10220.

1990s, when experts predicted widespread computer failures as soon as computer dates switched to the year 2000.[12] Moreover, if cyberattacks can cause effects that seem like accidents, policymakers and operators may not be able to easily distinguish between the two.[13] One can only imagine what conclusions and decisions could be made if the 2018 false missile alert in Hawaii occurred because of a system glitch instead of someone pushing the wrong button.[14] In the era of more frequent cyberattacks, misattribution of accidents in nuclear weapons systems carries substantial risks for nuclear stability as well.

While such accidents are not deterrable by definition, the United States has tried to adapt to the deliberate cyber threats by adjusting its deterrence strategy. The 2018 Nuclear Posture Review highlighted expanding cyber threats to NC3 and networked systems and noted that U.S. nuclear forces serve to deter nuclear and non-nuclear attacks alike.[15] In parallel, U.S. Cyber Command (CYBERCOM) devised a strategy to deter malicious cyber operations in general by focusing on the "defend forward" concept—the policy of hunting for cyber threats at their source and thus focusing on adversarial cyberspace.[16] After Russia's invasion of Ukraine, for instance, CYBERCOM has conducted numerous hunt-forward operations to ensure the resilience of Ukrainian and NATO networks against "a range of cyber capabilities" employed by Russia's military and intelligence.[17] The tension between deterring cyberattacks on U.S. nuclear systems and conducting operations in adversarial cyberspace through the "defend forward" posture creates additional questions for nuclear stability that diplomacy can address.

The use of bilateral diplomacy in upholding strategic stability and deterring adversaries aligns with the concept of "integrated deterrence," released by the U.S. Department of Defense later in 2022 as part of the National Defense Strategy and said to combine all instruments of national power to maintain U.S. security and deter foreign threats.[18] As such, it is important to assess in greater detail the views of scholars and practitioners on the utility of diplomacy for mitigating cyber risks to nuclear stability.

The following subsection outlines two categories of views on the role of cyber-nuclear diplomacy with Russia on strategic stability. The first category, focusing on cyber deterrence, views bilateral diplomacy as an element of the broader "defend forward" posture. The second category sees U.S.-Russian diplomacy as a means to negotiate mutually beneficial risk reduction measures that would mitigate deliberate and accidental cyber risks and prevent nuclear escalation. The latter category can be further broken down into U.S. and Russian perspectives on the negotiability of cyber risk reduction measures.

12   Wendin Smith, "Russia's Nuclear Arsenal: Why the Y2K Bug Didn't Bite," *Fletcher Forum of World Affairs* 24, no. 1 (Spring 2000): 191–97, http://www.jstor.org/stable/45289106.
13   Page O. Stoutland and Samantha Pitts-Kiefer, "Nuclear Weapons in the New Cyber Age," Nuclear Threat Initiative, September 2018, 16, https://www.nti.org/analysis/articles/nuclear-weapons-cyber-age/.
14   Amy B. Wang, "Hawaii missile alert: How one employee 'pushed the wrong button' and caused a wave of panic," *Washington Post*, January 14, 2018, https://www.washingtonpost.com/news/post-nation/wp/2018/01/14/hawaii-missile-alert-how-one-employee-pushed-the-wrong-button-and-caused-a-wave-of-panic/.
15   U.S. Department of Defense, *Nuclear Posture Review* (Washington, DC: February 2018), vii–xiii, https://media.defense.gov/2018/Feb/02/2001872886/-1/-1/1/2018-NUCLEAR-POSTURE-REVIEW-FINAL-REPORT.PDF.
16   "Summary: Department of Defense Cyber Strategy," Department of Defense.
17   "Posture statement of Gen. Paul M. Nakasone, commander, U.S. Cyber Command before the 117th Congress," U.S. Cyber Command, April 5, 2022, https://www.cybercom.mil/Media/News/Article/2989087/posture-statement-of-gen-paul-m-nakasone-commander-us-cyber-command-before-the/.
18   "Integrated Deterrence at Center of Upcoming National Defense Strategy," U.S. Department of Defense, March 4, 2022, https://www.defense.gov/News/News-Stories/Article/Article/2954945/integrated-deterrence-at-center-of-upcoming-national-defense-strategy/.

ROLE OF BILATERAL DIPLOMACY: THE "DEFEND FORWARD" CATEGORY

When discussing ways to deter cyber threats overall, including at the cyber-nuclear nexus, some scholars integrate bilateral diplomacy into an approach that combines the "defend forward" posture with improving domestic cyber resilience and communicating with allies. This approach is encapsulated in the "layered cyber deterrence" concept coined by the 2020 Cyberspace Solarium Commission.[19] The three layers—shaping norms through allied diplomacy, enhancing cybersecurity, and imposing costs through the "defend forward" posture—leave little room for discussing the role of bilateral risk reduction and negotiations within the national cyber strategy. Within this strategy, the Department of State's role lies primarily in allied diplomacy and international engagement, seeking to stabilize interactions in cyberspace through multilateral negotiations of confidence-building measures and norm-shaping.

Similarly, CYBERCOM strategist Emily Goldman focuses on diplomacy primarily as a means of norm-building among like-minded allies and the private sector, designed to shape the security environment and embrace active competition with U.S. adversaries in cyberspace.[20] This approach, paired with steps to bolster cybersecurity and "defend forward" in adversary cyberspace, aligns with Goldman and Richard Harknett's argument in 2016 that the cyber domain is an "offense-persistent" strategic environment in which defense can be sustained while in constant contact with the adversary.[21]

This line of thinking appears to deprioritize diplomacy with adversaries, choosing to rely much more so on defensive measures, allied cooperation, and threat of retaliation to deter cyber threats. The only application of such adversarial engagement in the Solarium Commission's report can be found in the recommendation to design a signaling strategy using public diplomacy and private communications "through mechanisms such as hotlines and other nonpublic channels."[22] As such, within the context of upholding strategic stability and mitigating cyber risks to nuclear weapons vis-à-vis Russia, this category of literature might not treat bilateral diplomacy as a useful standalone instrument.

ROLE OF BILATERAL DIPLOMACY: THE "RISK REDUCTION" CATEGORY

Scholars in the second category put greater emphasis on the utility of bilateral adversarial diplomacy in cyber-nuclear risk reduction. Since the United States and Russia kickstarted the bilateral cybersecurity dialogue in 2021, Western and Russian analysts have brainstormed a series of various confidence-building measures that the two countries could negotiate to mitigate cyber threats to nuclear systems and thus threats to strategic stability overall. For instance, the Euro-Atlantic Security Leadership Group urged the two countries in June 2021 to adopt "cyber rules of the road" that could outline which parts of nuclear weapons enterprises are off-limits to cyber interference.[23] At the start of this year, the Young Deep Cuts Commission, consisting of rising American, German, and Russian

19   Cyberspace Solarium Commission, *Cyberspace Solarium Commission* (Washington, DC: March 2020), https://www.solarium.gov/report.

20   Emily Goldman, "Cyber Diplomacy for Strategic Competition," Foreign Service Journal, June 2021, https://afsa.org/cyber-diplomacy-strategic-competition.

21   Richard Harknett and Emily Goldman, "The Search for Cyber Fundamentals," *Journal of Information Warfare*, 15, no. 2 (2016): 86, https://www.researchgate.net/publication/305701575_The_Search_for_Cyber_Fundamentals.

22   Cyberspace Solarium Commission, *Cyberspace Solarium Commission*, 34.

23   "Advancing Strategic Stability in the Euro-Atlantic Region," Euro-Atlantic Security Leadership Group, June 2021, 4, https://www.europeanleadershipnetwork.org/wp-content/uploads/2021/06/EASLG-Statement_3_6.3.21.pdf.

scholars, gave its own recommendations for mitigating cyber-nuclear threats, listing confidence-building measures such as NC3 noninterference and doctrine transparency as examples of both bilateral and multilateral solutions.[24]

These proposals for bilateral diplomatic engagement track closely with U.S.-based writing on cyber deterrence through diplomacy. Joseph Nye argued in 2017 that the traditional dichotomy of deterrence by denial or punishment is not sufficient in cyberspace and that much more attention needs to be paid to deterrence by entanglement and norms.[25] These two mechanisms ostensibly work by raising costs through bilateral interdependence and mutually agreed-upon taboos of behavior in cyberspace. Other scholars are skeptical about the negotiability potential of multilateral norms and bilateral agreements but nevertheless recommend diplomacy in parallel with cybersecurity improvements as the least costly and escalatory way to deter cyber-nuclear risks.[26] And despite the difficulties of negotiating a full-scale cyber treaty with Russia even before the war in Ukraine, one proposal to do just that argued in 2019 that it is in the U.S. national interest to "negotiate some limits to this activity to reduce [cyber] threats."[27] These and other U.S.-based works on the value of adversarial diplomacy to reduce cyber risks to nuclear stability have outlined a valuable list of bilateral confidence-building measures to advance this goal over the last few years.[28]

Many Russian experts on cyber issues and diplomacy agree with the need to outline both countries' priorities, red lines, and shared definitions with regard to cyber-nuclear threats. Academics have previously advocated for the principle of noninterference, with NC3 as one of the most immediate and feasible solutions to this issue.[29] And yet, there are a few key differences that may affect Moscow's official stance on the prospect of engaging with the United States on this topic. For one, the Russian leadership has a public policy of refusing to negotiate on military applications of cyber technologies, and this policy would pose difficulties to discussing cyber-nuclear issues even before the war in Ukraine.[30] Furthermore, while the United States prioritizes deterring cyber threats through its retaliatory "defend forward" posture, Russia focuses on bolstering the cybersecurity of its critical systems and thus denying possible cyberattacks.[31] These differences in official perspectives make bilateral diplomacy a more difficult avenue of risk reduction and deterrence to pursue, especially with regard to the highly sensitive topics at the intersection of cyber and nuclear issues.

---

24   Grace Kier, Lindsay Rand, and Tim Theis, "'Hacking' Away at Risks at the Cyber- Nuclear Nexus," Young Deep Cuts, January 2022, 9, https://deepcuts.org/images/images/YDCC/YoungDeepCuts_PB4.pdf.

25   Joseph Nye, "Deterrence and Dissuasion in Cyberspace," *International Security* 41, no. 3 (2017), doi:10.1162/ISEC_a_00266.

26   Beyza Unal and Yasmin Afina, "How to Deter Cyberattacks on Nuclear Weapons Systems," Chatham House, December 18, 2020, https://www.chathamhouse.org/2020/12/how-deter-cyberattacks-nuclear-weapons-systems.

27   Robert G. Papp, "A Cyber Treaty with Russia," Wilson Center, March 2019, 2, https://www.wilsoncenter.org/sites/default/files/media/documents/publication/kennan_cable_no._41.pdf.

28   Erica Lonergan and Keren Yarhi-Milo, "Cyber Signaling and Nuclear Deterrence: Implications for the Ukraine Crisis," War on the Rocks, April 21, 2022, https://warontherocks.com/2022/04/cyber-signaling-and-nuclear-deterrence-implica-tions-for-the-ukraine-crisis/; and Jonathan Lancelot, "Cyber-diplomacy: cyberwarfare and the rules of engagement," *Journal of Cyber Security Technology* 4, no. 4 (2020), doi:10.1080/23742917.2020.1798155.

29   Alexey Arbatov, "Global Stability in the Nuclear World," *Russian Academy of Sciences Courier* 91, no. 6 (2021): 569, https://www.imemo.ru/files/File/ru/Articles/2021/VestnikRAS-062021-Arbatov.pdf.

30   Pavel Sharikov, "Prospects for US-Russia Cyber Rules of the Road: A Russian Perspective," Russia Matters, June 10, 2021, 53, https://russiamatters.org/sites/default/files/media/files/PDF-CyberRulesoftheRoad-061021-RMPaper.pdf.

31   Oleg Shakirov, "Whoever Comes with the Cyber-sword: Russian and US Approaches to Deterrence in Cyberspace," *Journal of International Analytics* 11, no. 4 (2020): 167, https://www.interanalytics.org/jour/article/view/326.

ARGUMENT

The previous section has demonstrated the distinct views on the utility of bilateral diplomacy for maintaining nuclear stability and deterring cyber operations on the United States' nuclear weapons enterprise. Notably, the two categories appear to focus on two complementing areas of diplomacy. The "defend forward" approach prioritizes the signaling role of adversarial diplomacy, thereby highlighting the *process* of diplomacy itself as a channel of deterrence messaging. Meanwhile, the risk reduction approach focuses on the overall context of strategic stability but seeks to address it chiefly through the *results* of diplomacy, placing greater premiums on negotiated outcomes. Each of these approaches thus emphasizes complementing pieces of adversarial diplomacy, and the highest potential of preserving strategic stability lies in the ability to synthesize these elements, balancing deterrence communication with efforts to negotiate stabilizing agreements.

In her seminal work *Diplomacy by Deterrence*, Anne Sartori views the process of diplomacy not as useless "cheap talk" but as an important mechanism to deliver threats and commitments through a variety of official and informal channels.[32] This mechanism became a critical part of nuclear deterrence at the onset of the Cold War. After the first use of nuclear weapons in Hiroshima and Nagasaki, first possessors of nuclear weapons held initial discussions about the catastrophic risk of nuclear annihilation that eventually led to arms control negotiations and formal regimes. Through this diplomatic process, the United States and the Soviet Union exchanged deterrence signals and developed a shared understanding of the nuclear danger, which ultimately contributed to restraint and nuclear stability.

It appears that cyber threats to nuclear weapons enterprises, novel in their technological nature and exacerbated further by the ongoing modernization of these arsenals' infrastructures, share a similarly unexplored and ambiguous nature. As such, this paper hypothesizes that the process of bilateral cyber-nuclear diplomacy enables a shared understanding of associated risks and thus bolsters strategic stability in and of itself. Specifically, this shared understanding may provide a channel for deterrence messaging and lay the foundation for formally negotiated restraints.

To evaluate this argument, this paper will first draw a parallel between the history of nuclear learning between the United States and the Soviet Union during the Cold War and the need for cyber-nuclear learning in the present day. Then, it will share the results of a set of expert interviews with scholars and practitioners in the field that will explore the utility of bilateral diplomatic channels.

EVIDENCE

NUCLEAR LEARNING IN THE COLD WAR

The nascent strategy of nuclear deterrence rapidly evolved as nuclear weapon states acquired new theoretical and conceptual knowledge about the bombs. In a process that Joseph Nye called "nuclear learning," the United States and the Soviet Union obtained and exchanged information that led to a shared understanding of the destructive power of nuclear weapons, escalation and proliferation risks, and dangers of arms racing.[33] Through this process and beyond, diplomacy has been a key element of shaping this understanding, ensuring that the other side also shares it, and delivering nuclear deterrence signals that rely on this common knowledge.

---

32   Sartori, Deterrence by Diplomacy, 5.

33   Joseph S. Nye, Jr., "Nuclear learning and U.S.-Soviet security regimes," *International Organization* 41, no. 3 (Summer 1987): 383, https://www.jstor.org/stable/2706750.

One of the early examples of diplomacy as a means to instill common knowledge was an effort to bridge conceptual gaps about the destructive nature of nuclear warfare. David Holloway detailed a thorough discussion at the 1955 Geneva Summit on the consequences of nuclear use and subsequent radioactive fallout between the leaders of the United States, the United Kingdom, and the Soviet Union that established a shared foundation of knowledge from which further learning could occur. In the words of the British prime minister Anthony Eden, "Each country present learnt that no country attending wanted war and each understood why . . . this situation had been created by the deterrent power of thermo-nuclear weapons."[34]

The "shared" component of knowledge—the notion that each side possessed the same knowledge—became a crucial element of deterrence and eventual restraint during the nuclear crises over West Berlin and the Soviet missile deployment in Cuba. In 1958, as Soviet leader Nikita Khrushchev threatened to challenge NATO's presence in West Berlin, U.S. president Dwight Eisenhower warned that a Soviet invasion of the city would lead to a nuclear war. In issuing this signal as a deterrent, Eisenhower relied on the knowledge that Khrushchev understood the catastrophic danger of a nuclear war in Europe, as well as on Khrushchev's ability to recognize that Eisenhower understood it just as well.[35] The past diplomatic experience at the Geneva Summit thus played an instrumental role in averting annihilation.

A similar dynamic, linking shared nuclear learning and subsequent restraint, emerged during the 1962 Cuban Missile Crisis between Khrushchev and President John F. Kennedy. As the Soviet Union sought to deter a U.S. invasion of Cuba, Khrushchev warned Kennedy against further escalation, writing that "you yourself understand perfectly of what terrible forces our countries dispose." Khrushchev also referred to this shared understanding when Fidel Castro proposed a preemptive nuclear strike against the United States if it invaded Cuba; the Soviet leader told Castro of the global destruction a thermonuclear war would entail and Kennedy's knowledge of that fact.[36] Once again, both adversaries were aware that they shared a base of knowledge about nuclear weapons that allowed them to exchange deterrent threats and avoid exercising them.

As the common foundation of nuclear learning became established, the United States and the Soviet Union became capable of bridging more complex conceptual gaps and engaging in a new mode of nuclear diplomacy: arms control. A key example is the shift toward mutual recognition that ballistic missile defense would never truly prevail over offense, leading to the eventual negotiation of the Anti-Ballistic Missile (ABM) Treaty. As both countries researched and developed missile defense systems, the United States was the first to recognize that the deployment of missile defenses would prove destabilizing by merely driving the adversary to pursue greater offensive developments. However, Defense Secretary Robert McNamara initially did not succeed in credibly sharing this view with his Soviet counterparts, who suspected instead that the United States had achieved a breakthrough and sought to prevent the Soviet Union from conducting further missile defense research before they acquired the same capability.[37]

---

34 David Holloway, "Nuclear Weapons and the Escalation of the Cold War, 1945–1962," in *The Cambridge History of the Cold War*, edited by Melvyn P. Leffler and Odd Arne Westad (Cambridge, UK: Cambridge University Press, 2010), 384.

35 Ibid., 392.

36 Ibid., 395.

37 David Holloway, "Racing toward Armageddon? Soviet Views of Strategic Nuclear War, 1955–1972," in *The Age of Hiroshima*, edited by Michael D. Gordin and G. John Ikenberry (Princeton, NJ: Princeton University Press, 2020), 83.

The two sides bridged this gap during the Strategic Arms Limitation Talks, resulting in the successful negotiation of the ABM Treaty. By the time the negotiators broached the topic, Soviet leadership became aware of the technical challenges associated with developing its own Avrora missile defense project and the futility of countering U.S. strategic capabilities.[38] This dialogue hinged on the recognition of the futility of developing a "genuinely impenetrable shield" and the messaging that emphasized the offensive deterrent capabilities. As McNamara himself noted in 1967, the United States had "already taken the necessary steps to guarantee that our offensive strategic weapons will be able to penetrate future, more advanced, Soviet defenses."[39] It seems fair to imagine that the same message was reiterated during ABM Treaty negotiations, leading both sides to share the knowledge that putting limits on missile defense was a key step toward strategic offensive arms control.

APPLICATION TO CYBER-NUCLEAR LEARNING

These examples from the Cold War demonstrate the utility of bilateral diplomacy, including formal negotiations, for developing a shared understanding of nuclear capabilities and risks that contributed to bilateral strategic stability. This understanding undergirded future deterrence messages and led to formal arms control outcomes. Crucially, in the case of the Geneva Summit, the diplomatic effort played an important role in delivering deterrence-related messages even without leading to any concrete agreements. History thus provides a rough model of cyber-nuclear deterrence and risk reduction in which adversarial dialogue between the United States and Russia becomes a substantial element of learning rather than just an intermediate step toward tangible agreements.

Of course, the model faces certain limitations in its application to today's world. Cold War nuclear diplomacy and strategic stability existed in a bipolar setting. In contrast, as Admiral Charles Richard, head of U.S. Strategic Command, pointed out, the United States now faces the challenge of deterring "two peer nuclear-capable opponents at the same time who have to be deterred differently."[40] For the United States, this entails the necessity to discuss strategic stability issues not just with Russia but with China as well. Therein lies the second key difference from the Cold War era: Washington and Beijing do not have the same experience of engaging in nuclear learning and developing formal control regimes. Cyber-nuclear issues may thus be secondary to the priority for U.S.-China engagement on fundamental nuclear risk reduction.

Still, cyber-nuclear risks remain the most pressing in the U.S.-Russian dyad because of the large sizes of their nuclear arsenals. And it may be impossible to stabilize and reduce cyber-nuclear risks for the two nations without conducting mutual learning on this topic. Past U.S.-Russian dialogues on cyberspace in 2013 and 2021 have focused on general confidence-building measures (including communications channels and incident notifications) and ransomware activity, respectively.[41] The two countries also played leading roles in the multilateral negotiations on Organization for Security and Cooperation in Europe cyber confidence-building measures and UN Group of Governmental Experts.

---

38   Ibid., 84.

39   "Text of McNamara Speech on Anti-China Missile Defense and U.S. Nuclear Strategy," *New York Times*, September 17, 1967, https://timesmachine.nytimes.com/timesmachine/1967/09/19/93873550.html?pageNumber=18.

40   Aaron Sanderford, "StratCom commander: U.S. adjusting to great power competition with Russia, China," *Nebraska Examiner*, July 28, 2022, https://nebraskaexaminer.com/2022/07/28/stratcom-commander-u-s-adjusting-to-great-power-competition-with-russia-china/.

41   "FACT SHEET: U.S.-Russian Cooperation on Information and Communications Technology Security," The White House, June 17, 2013, https://obamawhitehouse.archives.gov/the-press-office/2013/06/17/fact-sheet-us-russian-cooperation-information-and-communications-technol; and Olga Kiyan, "Establishing Cybersecurity Norms in the United Nations: The Role of U.S.-Russia Divergence," *Harvard International Review*, November 26, 2021, https://hir.harvard.edu/establishing-cybersecurity-norms-in-the-united-nations-the-role-of-u-s-russia-divergence/.

However, there is no evidence to suggest that the sides have discussed the cyber-nuclear nexus in depth. Even before the war in Ukraine, the two countries first had to discuss major differences between their approaches to cyber defense writ large, and the 2021 cybersecurity dialogue was explicitly focused on ransomware.[42] Just as in the lead-up to the ABM Treaty negotiations, there may be a conceptual gap in the recognition of miscalculation and escalation risks pertaining to cyber operations in the adversary's nuclear weapons and communications systems.

How would intergovernmental diplomacy on cyber-nuclear risks build shared understanding and strengthen strategic stability? Much like in various Track 2 dialogues between both countries' nongovernmental experts, the United States and Russia could explore theoretical escalation pathways and crisis scenarios that could arise through cyber interference in nuclear weapons and communications systems.[43] This discussion could also focus on the numerous dynamics of the ongoing digitization of nuclear weapons systems that could contribute to instability, including conventional-nuclear entanglement of command and control systems, ambiguity between cyber espionage and cyberattacks, rate of attribution, and potential for miscalculation.

Undoubtedly, the discussion of offensive cyber operations and the potential vulnerabilities of nuclear systems in an official bilateral setting would give both sides greater transparency into each other's capabilities. Following Sartori, this kind of communication could also possess inherent value for deterring the use of these capabilities through the threat of retaliation. Based on the analogous history of nuclear deterrence during the Cold War, a shared knowledge base of the aforementioned risks could lead to future restraint and eventual risk reduction measures.

INTERVIEWS

To further explore the question of how useful the process of bilateral U.S.-Russian diplomacy could be for achieving mutual transparency and understanding of cyber-nuclear risks, the author conducted a series of nine interviews with cyber and nuclear policy experts based in the United States, Europe, and Russia, including a senior Western cyber official and a senior U.S. State Department official. In those interviews, six nongovernmental participants agreed to speak on the record, while other experts and government officials spoke on background. The participants were asked to share their views on the U.S. and Russian threat perceptions of cyber-nuclear risks, the benefits of the diplomatic process for understanding these risks and exchanging communications of red lines and deterrence messages, and the prospects for codifying such restraint into concrete agreements.

In line with this paper's hypothesis, all experts and officials agreed on the general utility of bilateral diplomacy and informal discussions for sharing mutual concerns and developing a joint understanding of cyber-nuclear risks, thus contributing to strategic stability. Naturally, their views diverged instead in the adjacent topics, varying in the assessments of U.S. and Russian threat perceptions and confidence in both countries' interests to pursue diplomatic communication on this issue. The discussions also raised doubts about the likelihood of successful diplomatic results, emphasizing obstacles in translating a shared knowledge of risks into actual deterrence messaging and restraint policies, as well as highlighted a key conceptual ambiguity about the engagement dynamics at the intersection of the cyber and nuclear spheres.

---

42   Shakirov, "Whoever Comes With the Cyber-sword," 167; and Kiyan, "Establishing Cybersecurity Norms in the United Na-
       tions."
43   "U.S.-Russia Cyber-Nuclear Weapons Dialogue," Nuclear Threat Initiative, https://www.nti.org/about/programs-projects/
       project/u-s-russia-cyber-nuclear-weapons-dialogue/; and Stoutland and Pitts-Kiefer, "Nuclear Weapons in the New Cyber
       Age."

Interview participants offered mixed and tentative assessments of cyber-nuclear threat perceptions among the U.S. and Russian policymakers and academic experts, primarily because of the lack of official evidence. Almost all participants highlighted the doubly secretive nature of the cyber-nuclear nexus, which combines two of the most highly classified areas in both governments. As a result, some participants extrapolated their views from historical evidence, while others drew on the experience of various Track 2 dialogues between U.S. and Russian experts.

On the policymaker level, outside experts generally considered both U.S. and Russian perceptions of cyber-nuclear risks to be relatively low. While U.S. government efforts such as the Solarium Commission are paying closer attention to cyber risks to NC3 systems, there is little indication of understanding future vulnerabilities that may emerge with nuclear modernization.[44] Similarly, the Russian government may be maintaining a certain degree of confidence about their weapons' cybersecurity, carrying over from the Soviet Union, which particularly focused on creating "robust systems" and protecting NC3 from outside interferences.[45] A number of respondents pointed to the similar vagueness within the 2018 Nuclear Posture Review and Russia's 2020 Basic Guidelines of State Policy on Nuclear Deterrence on the topic of cyber risks to networked nuclear systems.[46] That vagueness may serve as an indication that the U.S. and Russian assessments of such threats are "not very comprehensive."[47] Furthermore, both countries' launch-on-warning postures suggest that they currently view the risk of cyber counterforce operations, which are difficult to identify accurately in short decisionmaking timeframes, as lower than that of a bolt-out-of-the-blue attack.[48]

These evaluations align with official reactions to this topic. A senior Western cyber official pointed out that there has been no public evidence of cyberattacks on nuclear weapon systems yet, and so the cyber-nuclear threat environment is difficult to assess. But the nuclear weapon states overall may still feel secure.[49] And within the context of the war in Ukraine, the lack of lethality from cyber incidents decreases the urgency of the related issues for the United States in comparison to kinetic warfare in the current conflict.[50]

On the expert level, comparisons of U.S. and Russian views on cyber-nuclear risks varied. Respondents did assess the U.S. scholarly sphere to be more sophisticated in the level of awareness of potential threats and escalatory dangers. One of the key pieces of evidence for this claim is the growing number of publications on this topic in the United States, which Russian academia has not yet matched.[51] However, the interviewees familiar with U.S.-Russian Track 2 dialogues on the cyber-nuclear nexus gave conflicting assessments of both sides' familiarity with the associated risks. For instance, while one respondent shared on background an experience prior to 2022 in which Russian experts readily engaged with U.S. views on possible cyber-nuclear escalation dynamics, another expressed pessimism about the number of experts in both countries that fully understand both the

---

44   Erica Lonergan, interview with the author, June 21, 2022.

45   Pavel Podvig, interview with the author, June 30, 2022.

46   "Об Основах государственной политики Российской Федерации в области ядерного сдерживания" [Basic Guidelines of the Russian Federation's State Policy on Nuclear Deterrence], President of Russia, June 2, 2020, http://static.kremlin.ru/media/events/files/ru/IluTKhAiabLzOBjlfBSvu4q3bcl7AXd7.pdf.

47   Oleg Shakirov, interview with the author, July 15, 2022.

48   Herbert Lin, interview with the author, July 8, 2022.

49   Senior Western cyber official, interview with the author.

50   Senior U.S. State Department official, interview with the author.

51   Shakirov, interview.

cyber and nuclear sides of the picture.[52] Moreover, even as Russian experts began to share the U.S. experts' understanding of the risks prior to 2022, they may be sharing their policymakers' confidence about the Russian nuclear arsenal's protection.[53]

The presumably nascent assessments of cyber-nuclear threats at the official level and divergent perceptions among U.S. and Russian experts indicate that there is a significant need for developing a shared understanding of these risks and escalation dynamics across both levels. Thankfully, this is where direct diplomacy may serve as a useful tool for conducting cyber-nuclear learning and signaling each other about relevant capabilities.

BENEFITS AND ROLE OF DIRECT U.S.-RUSSIAN DIPLOMACY

Many respondents gave positive assessments of the potential value of U.S.-Russian official negotiations and informal discussions for shared cyber-nuclear learning. Multiple participants singled out transparency as a key benefit that could result from these talks and cover mutual expectations, operational procedures, internal decisionmaking processes, and possibly even cybersecurity measures.[54] Once Track 1.5 or 2 dialogues on the subject resume, relevant literature recently published in the United States could serve as a starting point for getting the expert and policymaker communities on the same page.[55] In this context, several experts explicitly praised the list of 16 critical infrastructure sectors that President Biden delivered to President Putin at the 2021 Geneva Summit. And, in line with this paper's argument, respondents generally agreed that such transparency would be a first step toward cyber-nuclear risk reduction and stability.

At the same time, U.S.-based interviewees shared varying views on the way that such diplomatic engagement should fit into the country's broader policy of defending against cyber-nuclear threats. For instance, Erica Lonergan emphasized the importance of U.S. declaratory policy "that stipulates what we actually see as off limits, not just for our adversaries, but also for our own behavior" and thus serves as the basis for diplomatic norm-setting.[56] Jacquelyn Schneider noted the obstacles to her previous proposal of a "strategic cyber No-First-Use" declaratory policy but agreed that the rules of the road established through direct diplomacy should be integrated with the "defend forward" posture and ongoing cybersecurity enhancements.[57] The point of complementing bilateral and multilateral discussions on cyber-nuclear risks with cybersecurity enhancements at home was particularly crucial during most of the interviews.[58]

Lastly, one key benefit of increased cyber-nuclear transparency could be the ability to jointly discuss and mitigate potential accidents. The precursor for such work goes back to the establishment of the Center for Year 2000 Strategic Stability, a joint U.S.-Russian project to prevent a possible nuclear accident caused by the aforementioned Y2K bug.[59] While cyber vulnerabilities and technical malfunctions today are harder to foresee than the advent of the year 2000, offering clarifying assurances to the adversary to avoid escalatory misperceptions would be a very helpful practice for strategic stability.

---

52 U.S.-based expert, interview with the author.
53 Ibid.
54 Lonergan, interview; Podvig, interview; and Jacquelyn Schneider, interview with the author, July 6, 2022.
55 Jason Healey, interview with the author, June 29, 2022.
56 Lonergan, interview.
57 Jacquelyn Schneider, "A Strategic Cyber No-First-Use Policy? Addressing the US Cyber Strategy Problem," *Washington Quarterly* 43, no. 2 (2020): 159–75, doi:10.1080/0163660X.2020.1770970; and Schneider, interview.
58 Senior Western cyber official, interview.
59 Bob Brewin, "U.S./Russian Y2K center to avoid nuclear exchange," CNN, March 4, 1999, http://www.cnn.com/TECH/computing/9903/04/ameruss.y2k.idg/; Smith, "Russia's Nuclear Arsenal: Why the Y2K Bug Didn't Bite"; and Shakirov, interview.

RECOMMENDATIONS FOR DIALOGUE ORGANIZATION

As a senior U.S. State Department official has noted, the ongoing war in Ukraine and overall
lack of trust between the United States and Russia mean that the U.S. government currently has
"no predisposition to have a conversation" on the topic of any cyber issues with Russia.[60] Still,
assuming that broader U.S.-Russian diplomacy may resume at some point after the end of the war,
the respondents offered a range of proposals for how exactly a bilateral cyber-nuclear dialogue
should be organized. Many emphasized Track 1.5 and 2 diplomacy as an important foundation for
building the shared understanding and engaging diverse groups of participants, including former
policymakers, military officials, and cyber and nuclear policy academics, among others. The key goal of
such endeavors would be to "cross pollinate both" of these fields and increase the number of officials
and nongovernmental experts that understand both sides of the cyber-nuclear nexus.[61] One possible
way to achieve this goal, brought up independently by several interviewees, could be to engage the
work done on cybersecurity of civilian nuclear power plants, thus incorporating efforts of NGOs such
as the EastWest Institute.[62]

Moving onto the official dimension of cyber-nuclear talks, the theme of cross-pollination became
even more relevant. Several respondents viewed nuclear arms control and risk reduction dialogue
as an easier venue for broaching cyber-nuclear risks, potentially starting as side conversations and
informal dinner discussions.[63] One person familiar with a past U.S.-Russian Track 2 cyber-nuclear
dialogue underscored the importance of including the space domain as an adjacent but equally
important dimension of any exploration of cyber-nuclear risks.[64] With the right scope and diversity
of participants, success in gaining shared expertise would also depend on having officials and
operators of comparable function and rank on both sides of the table.[65] It is important to bear in
mind here that the Russian military does not appear to have any structure parallel to the Pentagon's
Cyber Command.[66] Still, a joint institutional project on a similar level to lab-to-lab or International
Space Station collaboration efforts could further deepen cyber-nuclear learning and provide mutual
institutional awareness in this sphere.[67]

Finally, it is worth noting that a U.S.-Russian exchange in cyber-nuclear issues may potentially also
cover U.S. weapon systems stationed in Europe. Hence, if the United States and Russia engaged
in cyber-nuclear talks after the war in Ukraine, involvement of NATO allies would likely depend
on the United States and Russia.[68] Hypothetically, this engagement could be modeled on the
extensive consultations with NATO allies that helped the United States develop its approach to the
Intermediate-Range Nuclear Forces Treaty.[69]

---

60   Senior State Department official, interview.

61   Lonergan, interview.

62   Senior Western cyber official, interview; and Bruce McConnell and Greg Austin, "A Measure of Restraint in Cyberspace,"
      EastWest Institute, January 31, 2014, https://www.eastwest.ngo/idea/measure-restraint-cyberspace.

63   Healey, interview.

64   U.S.-based expert, interview.

65   Podvig, interview.

66   Shakirov, interview.

67   Pavel Podvig, "U.S.-Russian Cooperation in Missile Defense: Is It Really Possible?," Russian Forces, April 15, 2003, https://
      russianforces.org/podvig/2003/04/us-russian_cooperation_in_miss.html.

68   Senior Western cyber official, interview.

69   "Treaty Between The United States Of America And The Union Of Soviet Socialist Republics On The Elimination Of Their
      Intermediate-Range And Shorter-Range Missiles (INF Treaty)," U.S. State Department, https://2009-2017.state.gov/t/avc/
      trty/102360.htm.

OBSTACLES TO ADOPTING CYBER-NUCLEAR RESTRAINTS

However, while all interviewees universally agreed on the stabilizing value of the process of cyber-nuclear diplomacy between the United States and Russia, most of them raised doubts about the prospect of credible signaling and restraint mechanisms that could emerge from such dialogue.

One of the central themes of this resistance focused on the principle of reciprocity; the United States should be willing to forswear its own use of certain practices that it would wish to put red lines around. Indeed, future cyber confidence-building measures could focus on refraining from what would be "the most destabilizing actions one might take in cyberspace" in order to promote strategic stability and reduce risk, including refraining from cyber-targeting NC3 systems. However, ascertaining these restraint measures could be extremely difficult, and if the act of forbearing a certain kind of cyberattack invites more questions than answers, it will be destabilizing.[70]

Hence, some U.S.-based interviewees immediately excluded certain practices that they could not realistically imagine the United States would forgo, including operations in peacetime to gain preemptive access into NC3 and nuclear weapons systems for espionage and wartime disruption purposes.[71] In this vein, the proposal of NC3 noninterference is complicated by the growing entanglement of nuclear and conventional command and control infrastructure and space-based systems.[72] Additionally, it would force both countries to delineate which systems they consider to be off-limits and thus potentially clash with the "transparency-security trade-off," especially when considering the growing entanglement of nuclear and conventional command and control infrastructures.[73] As Jacquelyn Schneider discovered in her wargaming research, another obstacle is the likelihood that policymakers might consider "a [cyber] counterforce option that is less escalatory than sending bombers" a good thing.[74] Perhaps most crucially, it is inherently difficult to verify that a given actor is fulfilling the promise to not conduct certain cyber operations.[75] The potential use of advanced persistent threats raises the difficulty of such verification even further.

Without a clear set of operations that the United States might wish to forswear at the cyber-nuclear nexus, deterrence messaging becomes a much harder endeavor as well. First, as one official noted, "signaling can take place at many different levels and in many different ways," including in private discussions, but the sides need to be committed to those signals in order to avoid miscalculated escalation.[76] Second, the absence of public evidence of cyber operations on nuclear weapons systems and the (undoubtedly fortunate) lack of massively destructive events caused by cyber disruptions increase the challenge of translating shared understanding of cyber-nuclear risks into an urgency to reduce those risks.[77] This poses an additional challenge to the model of the U.S.-Soviet nuclear learning during the Cold War, when deterrence messaging built on mutual recognition of the scale of potential nuclear annihilation.

---

70   Senior State Department official, interview.
71   Lonergan, interview; Healey, interview; and Schneider, interview.
72   Expert familiar with U.S.-Russian cyber-nuclear Track 2 dialogues, interview.
73   Jane Vaynman, "Better Monitoring and Better Spying: The Implications of Emerging Technology for Arms Control," *Texas National Security Review* 4, no. 4 (Fall 2021), doi:10.26153/tsw/17498; and Schneider, interview.
74   Schneider, interview.
75   Lin, interview.
76   Senior Western cyber official, interview.
77   Ibid.; and Podvig, interview.

Finally, while outside the planned scope of the interviews, several respondents raised an important conceptual point that warrants further exploration: the question of framing cyber-nuclear operations through the lenses of "deterrence" or "persistent engagement." In line with past writing on this subject by cyber strategists, the perception of the cyber domain as "always on" creates difficulties for translating the dynamic of nuclear deterrence into the cyber domain because it is a permanently contested environment, with blurred lines between peace and wartime.[78] It may well be possible that "persistent engagement" and the U.S. strategy of "forward defense" incorporate competitors' nuclear weapons-related networks as well, and vice versa.[79]

However, this does not mean that the deterrence frame never applies to specific cyber actions that could destabilize nuclear weapons systems. While most respondents cited the lack of public evidence as a primary obstacle to resolving this debate, several noted a "strategic tension" within official doctrines and rhetoric between the practice of persistent cyber operations below the level of armed conflict and the threat of nuclear response to certain unspecified kinds of cyberattacks.[80] Additionally, as Jason Healey observed, within the United States itself there is a separation between cautious cybersecurity and national security experts, who urge restraint from cyber operations on nuclear systems, and the community of cyber and nuclear warfighters, whose primary objective is to find ways to prevail over potential adversaries.[81] These tensions indicate that the conceptual questions of framing the cyber-nuclear nexus exist not just at the academic level but most likely within the policymaking structures of nuclear weapon states themselves.

CONCLUSION

This paper examined the value of bilateral diplomacy between the United States and Russia for discussing, deterring, and stabilizing the complex web of cyber-nuclear risks. To that end, it drew on historical evidence of U.S.-Soviet nuclear learning and hypothesized that the development of a shared understanding of cyber-nuclear risks could contribute to strategic stability, in part by incentivizing both countries to restrain each other's activities at this nexus and lending credibility to the associated deterrence messages. To explore this hypothesis, the study included a series of interviews with experts and officials involved in U.S. and Russian nuclear and cyber policies. The interviews partially supported the hypothesis; the respondents generally agreed on the stabilizing value of the diplomatic process to accelerate mutual understanding of cyber-nuclear risks. They did, however, express skepticism about the prospects of turning that understanding into deterrence signals or negotiated restraints.

Reading closely the separate themes that emerged from the interviews, the experts' assessments of U.S. and Russian threat perceptions at the cyber-nuclear nexus mapped onto the nuanced differences that certain Russian experts already tracked in the existing literature. Meanwhile, the consideration of the technical and practical challenges for negotiating reciprocal restraints further underscores the difficulty of expecting agreed-upon measures of cyber-nuclear risk reduction between the United States and Russia in the near-to-medium term. Perhaps more importantly, these challenges pale in comparison to fundamental conceptual issues at the intersection of cyber engagement and nuclear deterrence relationships that remain unresolved and need to be tackled in future research.

---

78   Harknett and Goldman, "The Search for Cyber Fundamentals"; and senior Western cyber official, interview.

79   Lin, interview.

80   Lonergan, interview.

81   Healey, interview.

Despite these obstacles, the overwhelmingly positive responses to the value of cyber-nuclear dialogue indicate that the United States should pursue discussions on this topic once Russia ceases its invasion of Ukraine and both governments resume their interactions on strategic nuclear issues. Even if the interviewees' intuition bears out, and concrete restraint measures and credible deterrence messages remain out of reach, the cyber-nuclear learning will carry an inherent benefit for strategic stability by increasing mutual transparency and clarity. Therefore, it remains imperative to address the inadvertent and deliberate risks of nuclear escalation, once more brought into stark public view by the Russian invasion of Ukraine and include the cyber component of those risks in the discussion.

The dialogue's role in developing a common knowledge of cyber-nuclear risks can also assuage the misgivings about transforming that knowledge into deterrence communication. An understanding of escalation risks within the cyber-nuclear nexus that is shared across nations, expert communities, and distinct policy areas might contribute to resolving conceptual gaps and equalizing threat perceptions. Just like after the 1955 Geneva Summit, this common learning could sharpen and clarify deterrence communication that is carried out through formal diplomatic channels, military-to-military and cyber official hotlines, and the highest levels of governments. After all, the war in Ukraine has demonstrated that Washington still has the muscle memory to send deterrence messages to Moscow amid the efforts to deter a Russian nuclear use in the conflict.[82] In that sense, cyber-nuclear dialogue with Russia would directly follow from the United States' 2022 doctrine that links nuclear deterrence and risk reduction into an integrated approach.[83]

And while it is currently difficult to predict how said cyber-nuclear learning and deterrence communication could translate into future risk reduction measures, the Cold War experience also shows the link between these processes. Current assessments of what operations the U.S. government would or would not be willing to forswear within the cyber-nuclear space are not set in stone, and a better understanding of the risks associated with those operations might alter the calculus. Much like in the lead-up to the SALT I discussions, mutual cyber-nuclear learning could affect policymakers' will to refrain from certain kinds of strategic cyberattacks if they deem them to be significantly destabilizing. Joint work in this sphere would be key to figuring out the best verifiable measures through which nuclear weapon states can commit to existing proposals such as noninterference with NC3 and mutual vulnerability disclosures or come up with entirely new methods.

Moreover, considering the spread of cyber capabilities around the world, these discussions need not be limited to the U.S.-Russian dyad. Although it has been noted previously that cyber-nuclear risks are secondary in the context of U.S.-Chinese nuclear risk reduction, it may be possible to incorporate these issues into the nuclear dialogue if and when Washington and Beijing agree to it. Similar discussions can be held with other nuclear weapon states, both in bilateral and multilateral settings, in order to develop and shape international norms of responsible behavior at the cyber-nuclear nexus. While multilateral diplomacy lies outside the scope of this paper, it is worth recalling Joseph Nye, Jr.'s argument that these norms could feed back into deterrence communication and prove effective in deterring states that care about their global reputations from conducting destabilizing cyber operations.[84]

---

82   Paul Sonne and John Hudson, "U.S. has sent private warnings to Russia against using a nuclear weapon," *Washington Post*, September 22, 2022, https://www.washingtonpost.com/national-security/2022/09/22/russia-nuclear-threat-us-options/.

83   "Fact Sheet: 2022 Nuclear Posture Review and Missile Defense Review," U.S. Department of Defense, March 29, 2022, https://media.defense.gov/2022/Mar/29/2002965339/-1/-1/1/FACT-SHEET-2022-NUCLEAR-POSTURE-REVIEW-AND-MIS-SILE-DEFENSE-REVIEW.PDF.

84   Nye, "Deterrence and Dissuasion in Cyberspace."

Nevertheless, obstacles to effective deterrence communication and negotiated restraints in the cyber-nuclear nexus remain significant and will require much creativity to overcome. Due to the myriad of technical and conceptual difficulties related to cyber deterrence and verification, these effects of bilateral adversarial diplomacy may not bear out. But the very process of direct diplomacy and the act of mutual cyber-nuclear learning can still improve strategic stability. The United States should therefore include this topic in future nuclear talks with Russia after the Kremlin ceases its unbridled aggression in Ukraine. For although the war will end, the nuclear shadow hanging over the world will not, and it will be crucial to continue using all available measures to avoid catastrophe.

About the Editors

Reja Younis is the associate fellow with the Project on Nuclear Issues in the International Security Program at the Center for Strategic and International Studies (CSIS). She is also a PhD student at the Johns Hopkins University School of Advanced International Studies and a predoctoral fellow with the Henry A. Kissinger Center for Global Affairs. At CSIS, she leads research on nuclear deterrence issues, nuclear strategy, and emerging technologies. Prior to working at CSIS, she completed a year-long fellowship with the Stimson Center, where she conducted research on nuclear deterrence and escalation in the context of South Asia. Reja holds a BS in social sciences and liberal arts from the Institute of Business Administration and graduated with highest honors in political science. She completed her MA in international relations from the University of Chicago.

Jessica Link is a program coordinator and research assistant with the Project on Nuclear Issues in the International Security Program at the Center for the Strategic and International Studies (CSIS). Prior to joining CSIS, she was a research intern at the Wisconsin Project on Nuclear Arms Control. Jessica graduated from the College of William & Mary with a BA in government.